Lecture Notes in Netw

Volume 105

Series Editor

Janusz Kacprzyk, Systems Research Institute, Polish Academy of Sciences, Warsaw, Poland

Advisory Editors

Fernando Gomide, Department of Computer Engineering and Automation—DCA, School of Electrical and Computer Engineering—FEEC, University of Campinas—UNICAMP, São Paulo, Brazil

Okyay Kaynak, Department of Electrical and Electronic Engineering, Bogazici University, Istanbul, Turkey

Derong Liu, Department of Electrical and Computer Engineering, University of Illinois at Chicago, Chicago, USA; Institute of Automation, Chinese Academy of Sciences, Beijing, China

Witold Pedrycz, Department of Electrical and Computer Engineering, University of Alberta, Alberta, Canada; Systems Research Institute, Polish Academy of Sciences, Warsaw, Poland

Marios M. Polycarpou, Department of Electrical and Computer Engineering, KIOS Research Center for Intelligent Systems and Networks, University of Cyprus, Nicosia, Cyprus

Imre J. Rudas, Óbuda University, Budapest, Hungary

Jun Wang, Department of Computer Science, City University of Hong Kong, Kowloon, Hong Kong

The series "Lecture Notes in Networks and Systems" publishes the latest developments in Networks and Systems—quickly, informally and with high quality. Original research reported in proceedings and post-proceedings represents the core of LNNS.

Volumes published in LNNS embrace all aspects and subfields of, as well as new challenges in, Networks and Systems.

The series contains proceedings and edited volumes in systems and networks, spanning the areas of Cyber-Physical Systems, Autonomous Systems, Sensor Networks, Control Systems, Energy Systems, Automotive Systems, Biological Systems, Vehicular Networking and Connected Vehicles, Aerospace Systems, Automation, Manufacturing, Smart Grids, Nonlinear Systems, Power Systems, Robotics, Social Systems, Economic Systems and other. Of particular value to both the contributors and the readership are the short publication timeframe and the world-wide distribution and exposure which enable both a wide and rapid dissemination of research output.

The series covers the theory, applications, and perspectives on the state of the art and future developments relevant to systems and networks, decision making, control, complex processes and related areas, as embedded in the fields of interdisciplinary and applied sciences, engineering, computer science, physics, economics, social, and life sciences, as well as the paradigms and methodologies behind them.

** Indexing: The books of this series are submitted to ISI Proceedings, SCOPUS, Google Scholar and Springerlink **

More information about this series at http://www.springer.com/series/15179

Jinan Fiaidhi · Debnath Bhattacharyya ·
N. Thirupathi Rao
Editors

Smart Technologies in Data Science and Communication

Proceedings of SMART-DSC 2019

Editors
Jinan Fiaidhi
Department of Computer Science
Lakehead University
Thunder Bay, ON, Canada

N. Thirupathi Rao
Department of Computer Science
and Engineering
Vignan's Institute of Information
Technology
Visakhapatnam, Andhra Pradesh, India

Debnath Bhattacharyya
Department of Computer Science
and Engineering
Vignan's Institute of Information
Technology
Visakhapatnam, Andhra Pradesh, India

ISSN 2367-3370   ISSN 2367-3389 (electronic)
Lecture Notes in Networks and Systems
ISBN 978-981-15-2406-6   ISBN 978-981-15-2407-3 (eBook)
https://doi.org/10.1007/978-981-15-2407-3

© Springer Nature Singapore Pte Ltd. 2020
This work is subject to copyright. All rights are reserved by the Publisher, whether the whole or part of the material is concerned, specifically the rights of translation, reprinting, reuse of illustrations, recitation, broadcasting, reproduction on microfilms or in any other physical way, and transmission or information storage and retrieval, electronic adaptation, computer software, or by similar or dissimilar methodology now known or hereafter developed.
The use of general descriptive names, registered names, trademarks, service marks, etc. in this publication does not imply, even in the absence of a specific statement, that such names are exempt from the relevant protective laws and regulations and therefore free for general use.
The publisher, the authors and the editors are safe to assume that the advice and information in this book are believed to be true and accurate at the date of publication. Neither the publisher nor the authors or the editors give a warranty, expressed or implied, with respect to the material contained herein or for any errors or omissions that may have been made. The publisher remains neutral with regard to jurisdictional claims in published maps and institutional affiliations.

This Springer imprint is published by the registered company Springer Nature Singapore Pte Ltd.
The registered company address is: 152 Beach Road, #21-01/04 Gateway East, Singapore 189721, Singapore

Conference Committee Members

Organizing Committee

General Chair

Sabah Mohammed, Lakehead University, Canada
Debnath Bhattacharyya, Vignan's Institute of Information Technology, India

Advisory Board

Diego Galar, Lulea University of Technology, Sweden
Petia Radeva, University of Barcelona, Barcelona, Spain
Philippe Fournier-Viger, Harbin Institute of Technology, Guangdong, China
Oscar Cordon, Digital University, University of Granada, Spain
Tarek Sobh, University of Bridgeport, Connecticut, USA
Richard G. Bush, Black Hawk College, USA
Susmit Shannigrahi, Tennessee Tech University, USA
Haeng-kon Kim, Catholic University of Daegu, Korea
Javier Garcia-Villalba, Complutense University of Madrid, Spain
L. Rathaiah, Vignan Group, India
Krishna Lavu D., Vignan Group, India
V. Madhusudhan Rao, VFSTR University, India
Pavan Krishna Kosaraju, Vignan's Institute of Information Technology, India

Editorial Board

Jinan Fiaidhi, Lakehead University, Canada
N. Thirupathi Rao, Vignan's Institute of Information Technology, India

Program Chair

Tai-hoon Kim, BJTU, China
Osvaldo Gervasi, Perugia University, Italy

Finance Committee

Mr. B. Dinesh Reddy, Vignan's Institute of Information Technology, India

Local Arrangements Committee

S. Nagamallik Raj, Vignan's Institute of Information Technology, India
E. Stephen Neal Joshua, Vignan's Institute of Information Technology, India

Technical Programme Committee

Sanjoy Kumar Saha, Jadavpur University, Kolkata
G. S. Tomar, THDC-IHET, India
Deepika Koundal, Chitkara University, India
Ranadhir Mukhopadhyay, NIO-Goa, India
Hans Werner, University of Munich, Munich, Germany
Goutam Saha, Scientist, CDAC, Kolkata, India
Samir Kumar Bandyopadhyay, University of Calcutta, India
Y. Byun, Jeju National University, Jeju Island, Republic of Korea
Alhad Kuwadekar, University of South Wales, UK
Bapi Gorain, LUC, KL, Malaysia
Poulami Das, Heritage Institute of Technology, Kolkata, India
Indra Kanta Maitra, BPPIMT, Kolkata, India
Divya Midhun Chakkaravarthy, LUC, KL, Malaysia
F. C. Morabito, Mediterranea University of Reggio Calabria, Italy
Bidyut Gupta, Southern Illinois University Carbondale, USA

Conference Committee Members

Nancy A. Bonner, University of Mary Hardin-Baylor, Belton, USA
Alfonsas Misevicius, Kaunas University of Technology, Lithuania
Ratul Bhattacharjee, AVP, AxiomSL, Singapore
Lunjin Lu, Oakland University, Rochester, USA
Ajay Deshpande, CTO, Rakya Technologies, Pune, India
Debasri Chakraborty, BIET, Suri, West Bengal, India
Bob Fisher, The University of Edinburgh, Scotland
Alexandra Branzan Albu, University of Victoria, Victoria, Canada
Maia Hoeberechts, Ocean Networks Canada, University of Victoria, Victoria, Canada
M. H. M. Krishna Prasad, UCEK, JNTUK Kakinada, India
Edward Ciaccio, Columbia University, New York, USA
Yang-sun Lee, Seokyeong University, South Korea
Yun-sik Son, Dongguk University, South Korea
Jae-geol Yim, Dongguk University, South Korea
Jung-yun Kim, Gachon University, South Korea
Mohammed Usman, King Khalid University, Abha, Saudi Arabia
Xiao-Zhi Gao, University of Eastern Finland, Finland
Tseren-Onolt Ishdorj, Mongolian University of Science and Technology, Mongolia
Khuder Altangerel, Mongolian University of Science and Technology, Mongolia
Jong-shin Lee, Chungnam National University, South Korea
Jun-kyu Park, Seoul University, South Korea
Wang Jin, Changsha University of Science and Technology, China
Goreti Marreiros, IPP/ISEP, Portugal
Mohamed Hamdi, Supcom, Tunisia

Preface

Knowledge in engineering sciences is about sharing our ideas of research to others. In engineering, it has many ways to exhibit. Among them, conference is the best way to propose your idea of research and its future scope, and it is the best way to add energy to build strong and innovative future. So, here we are to give a small support from our side to confer your ideas by an "International Conference on Smart Technologies in Data Science And Communication(SMART-DSC 2019)", related to electrical, electronics, information technology and computer science. It is not confined to a specific topic or region, and you can exhibit your ideas in similar or mixed or related technologies bloomed from anywhere around the world because "An idea can change the future and its implementation can build it". VIIT College is a great platform to make your idea(s) penetrate into the world. We give as best as we can in every aspect related. Our environment leads you to a path on your idea, our people will lead your confidence, and finally, we give our best to make yours. Our intention is to make intelligence in engineering to fly higher and higher. That is why we are dropping our completeness into the event. You can trust us on your confidentiality. Our review process is double-blinded through Easy Chair.

At last, we pay the highest regard to the Vignan's Institute of Information Technology, a "not-for-profit" Society from Guntur and Visakhapatnam for extending support for the financial management of 3rd SMART-DSC 2019.

Best wishes from:

Thunder Bay, ON, Canada	Prof. Jinan Fiaidhi
Visakhapatnam, India	Prof. Debnath Bhattacharyya
Visakhapatnam, India	Dr. N. Thirupathi Rao

Acknowledgements

The editors wish to extend heartfelt acknowledgement to all contributing authors, esteemed reviewers for their timely response, members of the various organizing committees and production staff whose diligent work put shape to the 3rd SMART-DSC 2019 proceedings. We especially thank our dedicated reviewers for their volunteering efforts to check the manuscript thoroughly to maintain the technical quality and for useful suggestions.

We also pay our best regards to the faculty members of Vignan's Institute of Information Technology for extending their enormous assistance during the conference-related assignments.

We also acknowledge the financial support received from our esteemed institute. At last, we extend our sincere thanks to Springer Nature for agreeing to be our publishing partner. Especially, the efforts made by Aninda Bose, Executive Editor, are highly appreciable.

Prof. Jinan Fiaidhi
Prof. Debnath Bhattacharyya
Dr. N. Thirupathi Rao

Contents

Digital Transformation of Seed Distribution Process 1
Talasila Bharat

Detection of Deceptive Phishing Based on Machine Learning
Techniques 13
J. Vijaya Chandra, Narasimham Challa and Sai Kiran Pasupuleti

A Shape-Based Model with Zone-Wise Hough Transformation
for Handwritten Digit Recognition 23
Dipankar Hazra and Debnath Bhattacharyya

Deducted Sentiment Analysis for Sarcastic Reviews Using LSTM
Networks 35
Labala Sarathchandra Kumar and Uppuluri Chaitanya

Automatic Identification of Colloid Cyst in Brain Through MRI/CT
Scan Images 45
D. Lavanaya, N. Thirupathi Rao, Debnath Bhattacharyya and Ming Chen

A Detailed Review on Big Data Analytics 53
Eswar Patnala, Rednam S. S. Jyothi, K. Asish Vardhan
and N. Thirupathi Rao

A Review on Datasets and Tools in the Research of Recommender
Systems 59
B. Dinesh Reddy, L. Sarath Chandra Kumar and Naresh Nelatur

Performance Comparison of Different Machine Learning Algorithms
for Risk Prediction and Diagnosis of Breast Cancer 71
Asmita Ray, Ming Chen and Yvette Gelogo

Analysis of DRA with Different Shapes for X-Band Applications 77
P. Suneetha, K. Srinivasa Naik, Pachiyannan Muthusamy and S. Aruna

Android-Based Application for Environmental Protection   85
Bonela Madhuri, Ch Sudhakar and N. Thirupathi Rao

LDA Topic Generalization on Museum Collections   91
Zeinab Shahbazi and Yung-Cheol Byun

Roof Edge Detection for Solar Panel Installation   99
Debapriya Hazra and Yung-Cheol Byun

Implementation of Kernel-Based DCT with Controller Unit 105
K. B. Sowmya, Neha Deshpande and Jose Alex Mathew

An Analysis of Twitter Users' Political Views Using Cross-Account
Data Mining 115
Shivram Ramkumar, Alexander Sosnkowski, David Coffman, Carol Fung
and Jason Levy

The Amalgamation of Machine Learning and LSTM Techniques
for Pharmacovigilance 123
S. Sagar Imambi, Venkata Naresh Mandhala and Md. Azma Naaz

An Artificial Intelligent Approach to User-Friendly Multi-flexible Bed
Cum Wheelchair Using Internet of Things 133
Bosubabu Sambana, Vurity Sridhar Patnaik and N. Thirupathi Rao

A Study on Pre-processing Techniques for Automated Skin
Cancer Detection 145
Netala Kavitha and Mamatha Vayelapelli

Prediction of Cricket Players Performance Using Machine
Learning 155
P. Aleemulla Khan, N. Thirupathi Rao and Debnath Bhattacharyya

Using K-means Clustering Algorithm with Python Programming
for Predicting Breast Cancer 163
Prasanna Priya Golagani, Shaik Khasim Beebi
and Tummala Sita Mahalakshmi

Compact Slot-Based Mimo Antenna for 5G Communication
Application 173
Sourav Roy, Srinivasa Naik, S. Aruna and S. K. Gousia Begam

DGS-Based Wideband Microstrip Antenna for UWB Applications 181
Y. Sukanya, Viyapu Umadevi, P. A. Nageswara Rao, Ashish Kumar
and Rudra Pratap Das

Brain Tumor Segmentation Using Fuzzy C-Means and Tumor Grade
Classification Using SVM 197
V Ramakrishna Sajja and Hemantha Kumar Kalluri

Optimized Water Scheduling Using IoT Sensor Data in Smart Farming 205
Kolli Venkatra Krishna Kishore, B. Yaswanth Kumar and S. Venkatramaphanikumar

Optimized Cylindrical and Rectangular DR Antenna for Ultra-Wideband Applications 221
K. Srinivasa Naik, D. Madhusudhan, S. Chandini and S. Aruna

A Stitch in Time Saves Nine: A Big Data Analytics Perspective 227
T. Archana Acharya and P. Veda Upasan

Flexi-Lexicon Learning Using Krill Herd Algorithm for Sentiment Analysis 245
Muddada Murali Krishna, JayaVani VanKara, V. Satyanarayana Kalahasthi and Ming Chen

Analysis of Queuing Model-Based Cloud Data Centers 251
K. V. Satyanarayana, K. Sudha, N. Thirupathi Rao and Ming Chen

Soft Computing and Big Data Intelligence for a Low-Carbon Economy 263
Jason Levy

Lung Image Classification to Identify Abnormal Cells Using Radial Basis Kernel Function of SVM 279
Sajja Tulasi Krishna and Hemantha Kumar Kalluri

Smart Technologies for Data-Driven Pipeline Risk Assessment 287
Jason Levy

Neuro-fuzzy Knowledge Processing for Smart Emergency Management: Advances in Computational Intelligence and Seismic Community Resilience 303
Jason Levy

Author Index 317

About the Editors

Dr. Jinan Fiaidhi is an internationally acclaimed author and keynote speaker in the field of collaborative learning and machine learning. Since her early Ph.D. work on the discovery of learning indicators in student programs back in 1983, she has established a pioneering line of research that builds on her expertise in this field, and her landmark achievements include Virtual Scenebeans for P2P collaboration, semantic MOOCS, crowdsourcing-based learning, and the use of IoT for learning. Dr. Fiaidhi has provided workshops and keynotes in over thirty countries. She firmly believes that collaborative learning should be established using a variety of enabling smart technologies, like ambient intelligence, calm computing, crowdsourcing, social media, and machine learning. She is a senior member of IEEE; Chair of Big Data for eHealth at the IEEE ComSoc eHealth TC, and Department Chair of Extreme Automation at IEEE IT Pro. She has published 4 books and more than 160 refereed articles.

Prof. Debnath Bhattacharyya is a Professor at the Computer Science and Engineering Department and Dean R&D at Vignan's Institute of Information Technology (Autonomous) in Visakhapatnam, India. Dr. Bhattacharyya is also an Adjunct Professor at VFSTR University, Guntur, an invited International Professor at Lincoln University College, Malaysia, and former Foreign Professor at the Department of Multimedia Engineering, Hannam University, South Korea. He received his Ph.D. (Tech., Computer Science and Engineering) from the University of Calcutta, Kolkata. He is a member of ACM, ACM SIGKDD, IEEE, life member of CSI, India, and senior member of IACSIT, Singapore as well as of IAENG, Hong Kong. He has published 176 Scopus indexed papers and 113 SCI / Web of Science indexed papers. His research interests include security engineering, pattern recognition, biometric authentication, multimodal biometric authentication, data mining, and image processing. In addition, he a reviewer for various international journals and has published 6 textbooks on computer science and engineering.

Dr. N. Thirupathi Rao is currently an Associate Professor at the Department of CSE, Vignan's Institute of Information Technology (Autonomous), Visakhapatnam, India. He completed his M.Tech. in Computer Science and Technology and Ph.D. in Computer Science and Engineering at Andhra University, Visakhapatnam, in 2010 and 2014, respectively. He is an editorial board member of the International Journal of Extreme Automation and Connectivity in Healthcare (IJEACH), and Associate Editor of the Journal of Statistical Computing and Algorithms.

Digital Transformation of Seed Distribution Process

Talasila Bharat

1 Introduction

India is a nation where 65% of the population know how to do the magic of transforming **Mud into Food**; it could be overwhelming in a country like India where arable land is 159.7 million hectares (394.6 million acres). India is the second largest country in the world where 58% of population's primary source is agriculture and sadly it is also the national catastrophe of farmers, for committing of suicides since the 1990s, often by drinking pesticides, -due to their inability to repay loans mostly taken from banks and NBFCs to **purchase expensive seeds and fertilizers**, which is often marketed by foreign MNCs or for not getting their insurances on their lands on time.

1.1 Problem Definition

- State seed farms in India have been set up in all the states for the increase and to multiply high-yielding varieties of seeds and for the distribution of those seeds to the farmers. This state seed farm policy also promotes 'Seed Village Scheme' to facilitate production and timely availability of seeds of the desired required crops/varieties at the local level. The states have also set up some special seed banks in the nontraditional areas for meeting the demand for seeds during natural calamities, shortfall in seed production and other emergency situations.

In farming, the selection of the right variety of farming seeds is extremely important to a farmer when it comes to the cultivation of a crop. Different seeds behave

---

T. Bharat (✉)
Department of Information Technology, Vignan's Institute of Information Technology, Visakhapatnam, Andhra Pradesh, India
e-mail: talasilabharat97@gmail.com

differently according to the soil, climate, irrigation and other inputs of an area. A kind of variety of seeds that gives a quality crop and a premium price in one regional place may be totally rejected and unfavorable in another region of place. This type of rejection may sometimes happen in and over other districts and even villages. Every farmer wants to get the maximum return for his produce, and for this, a prudent selection of the seed variety is vital.

Example We have a kind of variety of groundnut seed (SG-84); this seed grows very well in the loamy soils of Punjab. However, if it is planted in a dry state like Odisha, the seed will not grow. This does not mean that groundnut cannot be grown in Odisha. All the farmers in Orissa have to use a different type of seed. There is a variety of groundnut seed 'Jawan' that thrives in the climate and soils of Odisha and fetches a big price for farmers in the region. Similarly, in the same way, there are varieties of seeds in different regions for other crops that are grown only in specific regions. For details on the type of seed suitable for the region farmer live in, see Figs. 1 and 2.

Risk is an important part of agriculture. Agriculture in India is highly receptive to risks like droughts and floods. It is necessary to protect the farmers mainly from the risks related to weather conditions, market conditions, pests and diseases, etc. The major problem with the insurances is lack of insurance consciousness among farmers and lack of reinsurance support from professional reinsurer.

According to an article of Times Of India (TOI), responding to a question the agriculture minister of the state Purushottam Rupala told the Lok Sabha on March 28, 2017 that the continuous suicides and intermittent protests by the farmers all across the country have prompted the Supreme Court to ask why the much publicized Pradhan Mantri Fasal Bima Yojana (PMFBY) had not provided relief to the hapless farmers, and in the very issue, another main point to be noted is that most of the profit is being gained by the insurance companies only.

Year Production	Total Seed Production(Lakhqtls.)	Share of private sector
2003-04	132.27	47.48%
2004-05	140.51	45.02%
2005-06	148.18	46.80%
2006-07	194.31	41.00%

Fig. 1 National Seeds Corporation of India and State Farms Corporation of India [1, 2]. *Source* Compiled by Seeds Division of DAC

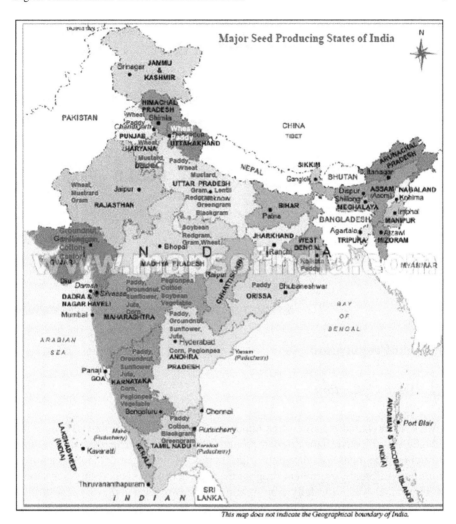

Fig. 2 Major seed-producing states are depicted in the map [3]

As per the pictorial representation, the insurance companies have collected 9081.8 crore rupees: from farmers (1643.3 Cr), central government (3708.7 Cr), state government (3729.9 Cr), for 2.5 crores of farmers, but only 32.7 lakhs of farmers benefited from it, and out of 2725.2 Cr, only 638.5 Cr was paid to the hapless farmers.

This shows the unfairness to the delay of insurance payments to the farmers even after 6 months after Kharif season 2016 ended. The insurance claims settled are a quarter of the total insurances of claims made. But there is a lot of scope for improvement and making an effective management tool (Fig. 3).

Fig. 3 *Source* Compiled by the Times of India (TOI) [3]

2 Data Preparation

2.1 Data Collection

After a research and meeting some hundreds of farmers from different regions and villages, we collected all the raw data related to their agriculture practices. We visited different government Web sites https://farmer.gov.in/ for gathering of information on crops and different government programmes and schemes, such as http://agricoop.nic.in/weather-watch for daily crop weather updates in each and every region from Crop Forecast Coordination Centre including seasons, reservoir status, fertilizer position, pest and diseases, availability position of seeds and fertilizers, All India Crop Situation and Progressive procurement of different crops, https://data.gov.in/sector/agriculture for collecting different datasets on all the information related to agriculture and http://agricoop.nic.in/ for all information related to important changes and agriculture in every region (Figs. 4, 5 and 6).

2.2 Data Preprocessing

Instead of state name and region name with its crops, we will categorize it according to the seeds and crops. By using the clustering algorithm, we decide how much seeds does each state need and estimate the profits to minimize the losses (Fig. 7).

METEOROLOGICAL SUB-DIVISIONWISE WEEKLY RAINFALL FORECAST & Wx. WARNINGS-2019								
Sr. No	MET.SUB-DIVISIONS	25 JUL	26 JUL	27 JUL	28 JUL	29 JUL	30 JUL	31 JUL
1	ANDAMAN & NICO.ISLANDS	FWS	FWS	FWS	WS	WS	WS	WS
2	ARUNACHAL PRADESH	WS*	FWS	FWS	WS*	WS*	WS*	FWS
3	ASSAM & MEGHALAYA	FWS*	FWS*	FWS	WS*	WS**	WS*	FWS
4	NAGA.MANI.MIZO.& TRIPURA	FWS*	FWS	FWS	WS*	WS*	WS*	FWS
5	SUB-HIM.W. BENG. & SIKKIM	WS**	FWS	FWS	FWS	FWS	WS*	FWS
6	GANGETIC WEST BENGAL	WS**	WS**	WS*	FWS	FWS	FWS	WS*
7	ODISHA	WS**	WS**	WS*	FWS	FWS	WS**	WS**
8	JHARKHAND	WS**	WS*	WS*	FWS	SCT	FWS	WS*
9	BIHAR	WS**	FWS	FWS	SCT	SCT	SCT	SCT
10	EAST UTTAR PRADESH	WS**	WS*	FWS	SCT	SCT	FWS*	WS*
11	WEST UTTAR PRADESH	WS**	WS*	FWS	SCT	SCT	FWS	WS*
12	UTTARAKHAND	WS**	WS**	FWS	FWS	FWS	WS*	WS**
13	HARYANA CHD. & DELHI	WS***	WS**	FWS	SCT	SCT	FWS*	WS*
14	PUNJAB	WS**	WS*	FWS	SCT	SCT	FWS*	WS*
15	HIMACHAL PRADESH	WS**	WS*	FWS	SCT	SCT	FWS	WS**
16	JAMMU & KASHMIR	WS*	WS*	FWS	SCT	SCT	SCT	FWS*
17	WEST RAJASTHAN	SCT**	SCT**	FWS**	FWS**	SCT	SCT	FWS
18	EAST RAJASTHAN	FWS**	WS**	WS**	WS**	FWS	FWS*	WS*
19	WEST MADHYA PRADESH	FWS**	WS**	WS**	WS*	FWS	FWS	WS**
20	EAST MADHYA PRADESH	WS**	WS**	WS**	WS*	FWS*	WS*	WS**
21	GUJARAT REGION D.D. & N.H.	SCT*	SCT*	FWS*	WS**	WS**	FWS	FWS*
22	SAURASTRA KUTCH & DIU	SCT	SCT	SCT	FWS**	WS**	FWS	FWS
23	KONKAN & GOA	WS**	WS**	WS**	WS**	WS**	WS*	WS**
24	MADHYA MAHARASHTRA	FWS**	FWS**	FWS*	FWS**	FWS**	FWS*	FWS*
25	MARATHAWADA	SCT	SCT	SCT	SCT	ISOL	ISOL	ISOL
26	VIDARBHA	FWS	FWS	WS*	FWS*	SCT	SCT	SCT
27	CHHATTISGARH	WS*	WS*	WS*	FWS**	FWS*	WS*	WS**
28	COASTAL A. PR. & YANAM	FWS*	FWS*	SCT	SCT	ISOL	SCT*	FWS*
29	TELANGANA	SCT	FWS	FWS	SCT	SCT	SCT	FWS*
30	RAYALASEEMA	SCT	ISOL	ISOL	ISOL	ISOL	ISOL	ISOL
31	TAMIL. PUDU. & KARAIKAL	SCT*	ISOL	ISOL	ISOL	ISOL	ISOL	ISOL
32	COASTAL KARNATAKA	WS*	WS*	WS*	WS*	WS*	WS	WS
33	NORTH INT.KARNATAKA	SCT	SCT	SCT	SCT	SCT	SCT	SCT
34	SOUTH INT.KARNATAKA	FWS	FWS	SCT	SCT	SCT	SCT	SCT
35	KERALA & MAHE	FWS	FWS	SCT	SCT	SCT	SCT	SCT
36	LAKSHADWEEP	WS	FWS	SCT	SCT	ISOL	ISOL	ISOL

LEGENDS:
WS – WIDE SPREAD / MOST PLACES (76-100%)   FWS – FAIRLY WIDE SPREAD / MANY PLACES (51% to 75%)
SCT – SCATTERED / FEW PLACES (26% to 50%)   ISOL – ISOLATED (up to 25%)   D/DRY – NIL RAINFALL
* Heavy Rainfall (64.5-115.5 mm)   ** Heavy to Very Heavy Rainfall (115.6-204.4 mm)   *** Extremely Heavy Rainfall (204.5 mm or more)
● FOG   * SNOWFALL   # HAILSTORM   ‖HEAT WAVE (+4.5 °C to +6.4 °C)   ‖ SEVERE HEAT WAVE (> +6.4)
$ THUNDERSTORM WITH SQUALL/GUSTY WIND   DUST/THUNDERSTORM   ‖COLD WAVE (-4.5 °C to -6.4 °C)   ‖SEVERE COLD WAVE (< -6.4)

Source: IMD

Fig. 4 Collection of data -1 [4]

2.3 Clustering

Crop productivity is based on multiple factors including type of seeds, soil, season, yield, etc. In order to increase the production, we have to produce seeds suitable for a particular area.

8. Progressive procurement of Wheat as on 05.07.2019

Table: 8.1

State	Target in marketing season 2019-20 (April - March)	Progressive Procurement as on 05.07.2019	
		In Marketing season 2019-2020	In Marketing season 2018-2019
Bihar	2.00	0.03	0.15
Haryana	85.00	93.20	87.37
Madhya Pradesh	75.00	67.25	69.67
Punjab	125.00	129.12	126.62
Rajasthan	17.00	14.11	15.32
Uttar Pradesh	50.00	37.00	52.94
Uttrakhand	2.00	0.42	1.10
All-India	357.00	341.33	353.69

Fig. 5 Collection of data [5]

Period	Storage as % of FRL	Storage as % of last year	Storage as % of 10 year's average level
Current Week	25	63	71
Last Week	24	32	28
The percentage to live capacity at FRL was 40% on 25/07/2018, 36% on 25/07/2017, 34% on 21/07/2016 and 37% on 23/07/2015.			

Source: CWC

There were 28 reservoirs having storage more than 80% of normal storage, 29 reservoirs having storage between 51% to 80%, 13 reservoirs having storage between 31% to 50% and 30 reservoirs having storage upto 30% of Normal Storage out of these 6 reservoirs having no live storage.

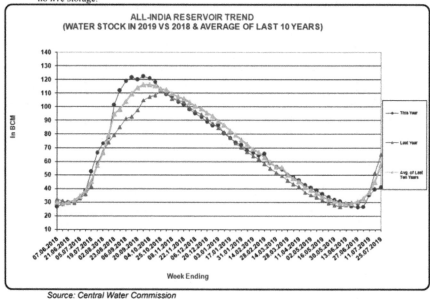

Source: Central Water Commission

Fig. 6 Water levels data [6]

Crop	State	Cost of Cultivation (`/Hectare) A2+FL	Cost of Cultivation (`/Hectare) C2	Cost of Production (`/Quintal) C2	Yield (Quintal/Hectare)
ARHAR	Uttar Pradesh	9794.05	23076.74	1941.55	9.83
ARHAR	Karnataka	10593.15	16528.68	2172.46	7.47
ARHAR	Gujarat	13468.82	19551.9	1898.3	9.59
ARHAR	Andhra Pradesh	17051.66	24171.65	3670.54	6.42
ARHAR	Maharashtra	17130.55	25270.26	2775.8	8.72
COTTON	Maharashtra	23711.44	33116.82	2539.47	12.69
COTTON	Punjab	29047.1	50828.83	2003.76	24.39
COTTON	Andhra Pradesh	29140.77	44756.72	2509.99	17.83
COTTON	Gujarat	29616.09	42070.44	2179.26	19.05
COTTON	Haryana	29918.97	44018.18	2127.35	19.9
GRAM	Rajasthan	8552.69	12610.85	1691.66	6.83
GRAM	Madhya Pradesh	9803.89	16873.17	1551.94	10.29
GRAM	Uttar Pradesh	12833.04	21618.43	1882.68	10.93
GRAM	Maharashtra	12985.95	18679.33	2277.68	8.05
GRAM	Andhra Pradesh	14421.98	26762.09	1559.04	16.69
GROUNDNUT	Karnataka	13647.1	17314.2	3484.01	4.71
GROUNDNUT	Andhra Pradesh	21229.01	30434.61	2554.91	11.97
GROUNDNUT	Tamil Nadu	22507.86	30393.66	2358	11.98
GROUNDNUT	Gujarat	22951.28	30114.45	1918.92	13.45
GROUNDNUT	Maharashtra	26078.66	32683.46	3207.35	9.33
MAIZE	Bihar	13513.92	19857.7	404.43	42.95
MAIZE	Karnataka	13792.85	20671.54	581.69	31.1

Fig. 7 Data processing [4, 5]

To produce seeds based on data analysis:

1. Gather information like production and supply from different datasets.
2. Aggregate the required data and form single data frame.
3. Clean the data.
4. Then, perform analysis on data to produce seeds in particular area using k-means clustering model (Fig. 8).

2.4 Algorithm

We use the k-means clustering algorithm and choose k as an arbitrary number, and instead of updating the k-means to be the centroid of each cluster, we set the new means to be the phrase which has the highest similarity in total with all the phases in the cluster.

Initialize k cluster centers
Do
Assignment step: Assign each point to its closest cluster center
Re-estimation step: Re-compute cluster centers
While (for still changes in cluster centers)

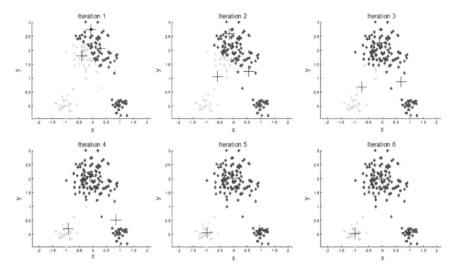

Fig. 8 Results

3 Model

3.1 Introduction to E-Smart Card/Agri Card

Agri E-Smart Card is a smart card just like credit/debit card, consists of RFID chip inside and has 13-digit unique number with a magnetic security strip on the backside for security. The smart card contains even the valid to expiry date where the user needs to change the card. The card contains the user's details, and it has a five-digit security pin. Unless the right pin is entered, the details will not be unlocked. It is a type of ledger (Fig. 9).

Fig. 9 Smart card model

3.2 Agri Smart Card Working Process

Every farmer gets a card from the government; this card works on a special machine upon scanning/swiping the card and asks for a five-digit pin. This pin is given to the farmers upon the card registration process. (Later, the pin can also be changed.) Upon entering the correct pin, the card gets unlocked; this Agri Smart Card contains all the farmers' details including the Aadhar card, passbook, crop insurances, bank details, all crop details, previous crop, previous fertilizers brought, previous government seed collection, subsidy seed information, etc. Depending upon the region, the climate in the area, dams, previous crop, etc., the Agri Smart Card suggests 4–5 options of seeds which can give a farmer maximum profit for a better growth; this information gathering process is done using the clustering method mentioned above.

This card holds all the details with just a swipe of a card, and these details of the cards are updated every time they make any purchase of seeds, fertilizers or claim insurances, etc. These data are sent to the local and central government to keep track of all the information from time to time.

3.3 Main Advantages of Smart Seed System with Agri Smart Card

Optimization of seed booking process for the farmers.
Optimization of subsidy seed procurement by farmers.
Optimization of seed delivery process by dealers to the farmers.
Ease of subsidy claim by dealers from the government.
Facilitation of crop insurance settlements in the case of crop damage.
Provision of purchasing crop insurance along with the purchase of subsidy seeds.

3.4 Smart Seed System with Agri Card—Implementation Phases

1. **Phase 1**: Agri Smart Card registration process (by the farmer): Farmer applies for the Agri Smart Card via local panchayat (Fig. 10).
2. **Phase 2**: Preseed booking process (by farmer for seed request): Farmer can get the information on what crops to invest to get maximum returns using the 15-digit pin on Agri E-Smart Card, which can be done via toll free/SMS/local gram panchayat. (The information on getting maximum returns from crops is gathered as mentioned in the above by considering various factors of water, previous crop, dam, weather, losses in the area, etc.)

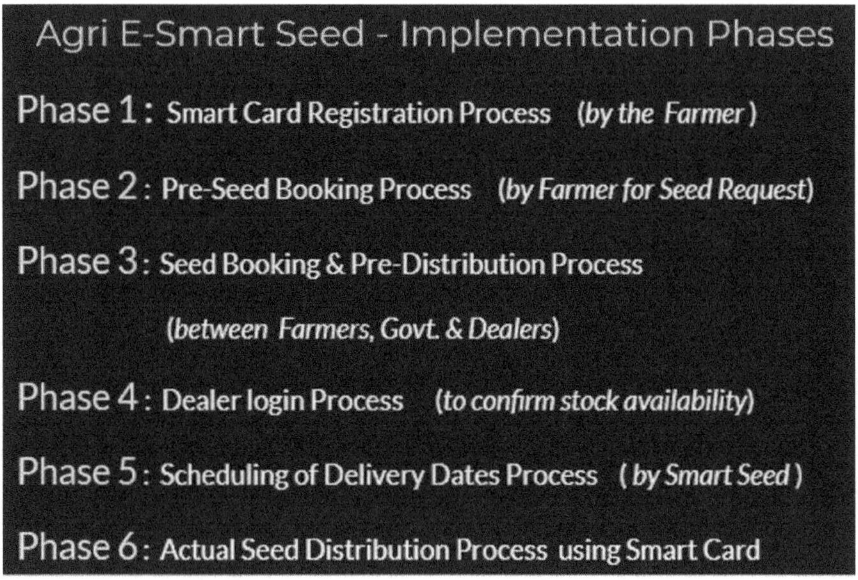

Agri E-Smart Seed - Implementation Phases

Phase 1: Smart Card Registration Process *(by the Farmer)*

Phase 2: Pre-Seed Booking Process *(by Farmer for Seed Request)*

Phase 3: Seed Booking & Pre-Distribution Process

  (between Farmers, Govt. & Dealers)

Phase 4: Dealer login Process *(to confirm stock availability)*

Phase 5: Scheduling of Delivery Dates Process *(by Smart Seed)*

Phase 6: Actual Seed Distribution Process using Smart Card

Fig. 10 Phases in the proposed model

3. **Phase 3**: Seed booking and predistribution process (between Farmers, Government and Dealers).): The farmers need to book the seeds 3–4 months prior to the distribution, and they need to mention the quantity and the type of seed they are requesting. Once booking confirmed, an SMS will be sent to the mobile for confirming and tracking their seeds. (They must mention 3–4 months prior so that the seed manufacturing company and the government will know the quantity needed to produce and send to an area.)
4. **Phase 4**: Dealer log-in process (to confirm stock availability): Upon receiving the stock in the region, the dealer confirms the stock availability.
5. **Phase 5**: Scheduling of delivery dates process (by smart seed): A specific date is allocated to the farmers to collect subsidy seeds; if a farmer misses to collect the seeds once, he can reschedule the process again within 30 days. (Rescheduling can be done only once for one season.)
6. **Phase 6**: Actual seed distribution process using smart card (to the farmer by dealer).

3.5 Insurance Claims Using Smart Seed System with Agri E-Card

The farmers can apply for insurance using their Agri Smart card by visiting their insurance companies. Once they apply for an insurance using the government Web site mentioned above, the insurance companies can verify whether really a natural calamity or anything happened and later send a verifier or a drone to take pictures of confirmation; once the government confirms an individual is eligible for an insurance claim, the individual can visit their insurance company to show their Agri E-Smart Card, which has all the details confirmed by the government. This is linked to a government Web site, which sends time-to-time updates to the government. Within 4–5 days of government's approval for a farmers' insurance eligibility, the insurance company will pay the farmer or else send the alert to the government and then an action will be taken.

4 Results and Conclusion

On given queries and gaps in current manual intervention and delay in insurance claims to farmers, our model has generated a plausible and logical way for the subsidy seed distribution. It also stores information on the crop details and farmers' details, and everything in it is locked with a five-digit pin.

For instance, a farmer wants subsidy seeds for the crop, takes the Agri Smart Card and enters the 13-digit unique number in the card into the phone via toll free or SMS or even by visiting their nearest panchayat. The algorithm considers all the factors and suggests the best 4–5 crops which could return maximum benefit. The farmer can choose one of the five or any seed he wishes and book a seed quantity 3–4 months prior. The farmer gets a Confirmation message and later a delivery date and even the rate of the seed is sent, with a provision to rescheduled only once for a season upon missing the collection date of Subsidy seeds.

Every time a farmer makes a purchase on fertilizers or seeds and claim for insurance or anything, the farmer must have this card to get anything and the details in the card get updated every time including the quantity, date, time and rate of the item. These data are even sent to the government from time to time to keep track of each and every status. The farmers can claim for insurance using this (Fig. 11).

A Small Sample Assistant Google Bot was created using Dialogflow to create awareness among the people of what is Agri Card and how it works.

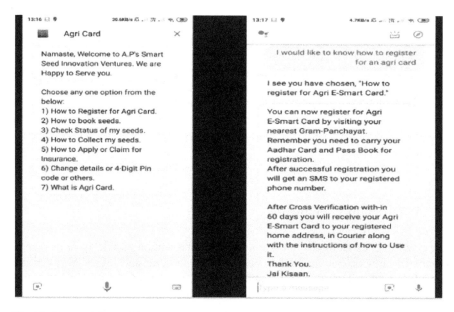

Fig. 11 Output of the model

References

1. History of Agriculture, https://en.wikipedia.org/wiki/Agriculture_in_India. Accessed 10-06-2019
2. Agriculture | National Portal of India, https://www.india.gov.in. Accessed 10-06-2019
3. Ministry of Agriculture & Farmers' Welfare, Government of India, https://www.agriculture.gov.in. Accessed 10-06-2019
4. Source: http://agricoop.nic.in/weather-watch, https://farmer.gov.in/, https://data.gov.in/sector/agriculture. Accessed 10-06-2019
5. Indian Meteorological Department (IMD), Central Water Commission (CMC). Accessed 10-06-2019

Image Sources

6. Source: Crops and TMOP Division, DAC&FW
7. Source: Fertilizer Division (DAC&FW)/Dept. of Fertilizer
8. Source: Compiled by Seeds Division of DAC
9. Source: Compiled by the Times of India (TOI)

Detection of Deceptive Phishing Based on Machine Learning Techniques

J. Vijaya Chandra, Narasimham Challa and Sai Kiran Pasupuleti

1 Introduction

Phishing attacks are the sophisticated targeted attacks done by an individual or a group with malicious intend for financial gain. The emails provide data transfer in the form of text, audio, video and multimedia communications, whereas websites are meant for sharing of information and money transactions. Information gathering is the preliminary task done by the malicious users by adopting different methods to extract information about legitimate users based on social engineering techniques. The attackers are targeting based on the information they obtained, and this information may be personal, professional or financial. The Internet is the fast-growing technology that connected the people all over the world virtually by removing all geographical barriers, due to fraudulent and malicious websites and spam emails, the people are losing millions of dollars, and the victims are increasing day by day. Phishing is a technique where a malicious user replicates trusted websites and sends spam emails with malicious uniform resource locators to deceive online legitimate users and finally gains access illegally using different techniques such as masking, spoofing, pharming, plugging and probing to steel financial assets [1].

Machine learning approach is a computing intensive and a subset of artificial intelligence which is commonly known as intelligent machine learning. It generally requires a large amount of training data or testing data. The machine learning and

---

J. V. Chandra (✉) · S. K. Pasupuleti
Department of Computer Science and Engineering, K. L. Deemed to be University, Koneru Lakshmaiah Education Foundation, Guntur, Andhra Pradesh, India
e-mail: vijayachandra.phd@gmail.com

S. K. Pasupuleti
e-mail: psaikiran@kluniversity.in

N. Challa
Department of Computer Science and Engineering, Vignan's Institute of Information Technology (A), Duvvada, Visakhapatnam, Andhra Pradesh, India
e-mail: drchallan@gmail.com

decision-making algorithms are implemented on datasets for classification, decision making and data analysis. Intelligent machine learning handles different approaches such as clustering, pattern recognition, regression analysis, classification and content-based detection for combating against phishing [2].

The goal is to combat with two tracks of phishing attacks, that are phishing emails attacks and phishing websites. Where first track phishing emails, attacks are done through emails where the victims are asked or provoked to click on the malicious URL or download the email attachment, and second track is phishing websites, where an attacker imitates official websites to mislead legitimate users and steals authentication credentials related to personal, official and financial system. These two tracks combinedly known as Deceptive Phishing. These two tracks combined known as Deceptive Phishing, where fraudsters send email messages which looks like legitimates possessions, claiming to originate from recognized sources as official correspondence are most frequently occurred fraudulent example is the PayPal Scammers where the email instructs to click on a link [3].

The detection and classification phishing using machine learning algorithms is based on the dataset methods on the anomaly spam classification, detection and filtering. The research is mainly focused on classifiers that are closely observed, to detect spam on the datasets, and good performance accuracy is done, moving toward combining classifier where the individual decisions of many classifiers are combined to get good classification [4].

2 Related Work

They are computing intensive that involves in repetitive training to improve the learning and decision making and generally implemented on large amount of datasets, by constructing predictive models. Machine learning explores construction of algorithms that can learn from data or datasets. Machine learning implements the mathematical, statistical and computational principles on the datasets. It adopts the continuous learning approach comes under the category of conditional probability from existing and new spams, and the email is divided into three parts that is body, header and attachments. The classifier generally prefers the body of the email for classification using the keywords and strings. These strings are part of the message and accountable, and the verification of email that is either spam or ham takes place based on tokens [5]. Combining classifiers or classification methods is possible for better performance and reduction of the error. Boosting is an ensemble technique that attempts to create a strong classifier from the number of weak classifiers. AdaBoost ensemble method for machine learning provides more accurate and efficient results. We import and extend the classifiers as done by using the Java.

3   Combining the Models of Machine Learning Algorithms

Boosting: The idea of the boosting is a continuous process that adds in weak learner to boost the data points that were not correctly classified. In order to solve complex and controversial datasets, we need more powerful model, where weaker learners need to boost, as weak learners do not have knowledge to classify or to complete the complicated task. If the predictions are wrong, it seems that the training mechanism using machine learning needs improvement, and such improvement process is possible with boosting. The combined model is more powerful, and the build model will improve the accuracy and efficiency. This model is also called as sequential learning model, where it combines the predictions or outputs which we get from the weak learners, and these weak learners are combined to get an improved accuracy, where it enhances the performance, the example of boosting algorithm is adaptive learning algorithm, and here the misclassified data points are identified and paid more attention toward classification of data points. It increases the weightage and new learners are added to improve classification; in order to get accuracy, even we update the weights as per the requirement based on the misclassification, a lot more data points are used to increase the efficiency. It generates multiple weak learners and generates rules for iterations. After the multiple iterations these rules are combined from multiple weak learners to make a strong rule set for classification.

Bagging: Bagging technique is also called as parallel ensemble learner which is used to incorporate the notation of bootstrapping and works using the principle of bootstrap sampling method. It trains number of weak learners in parallel, to improve the efficiency of ensemble model. Here we divide the dataset into different bootstrap datasets, the dataset is divided into different small subsets using parallel shifting methods and creates a classifier for each subset, and the results of these subsets are combined for the better results. The average results of these subsets classified results give better accuracy estimations compared with other methods.

Stacking: It performs better than the boosting and bagging. It uses meta-learner instead of voting to combine predictions of base learners, predictions of base learners are used as input for meta-learner, and base learners use different learning schemes. Base learners can output probabilities, and it uses predictions of multiple models as "features" to train a new model and uses the new model to make predictions on the test data. It is also called as generalization of the stack or stack generalization which involves in training the algorithm and involves in training an algorithm to combine the predictions of several other algorithms. The combiner algorithm is used to combine all predictions which are done by different algorithms, and this combiner algorithm gives a better result on available data.

Voting: It is the mechanism used for averaging of predictions of multiple pre-trained models, here predictions of each model are considered as a vote, the final prediction is most of predictions from the models, and hence, this final prediction is known as voting.

4 Machine Learning Methods

4.1 Holdout Method

In this method of supervised learning model, a model is trained using the labeled input data, to understand the performance of the model, the dataset is divided as training and testing data parts, this model is used to test the validation of the performance, in general the training data will be 70% as the test data will be 30% and division of data is done randomly to make sure that both buckets will have the similar data times, after the execution of the trained model the test model is compared for the validation and accuracy of the performance, and as it holds a part of input data for testing purpose, it is called a **holdout method**.

4.2 K-Cross-Validation Mechanism

K-cross-validation mechanism divides the dataset into K partitions, where data are distributed randomly in these partitions, these random portions are called as the folds as the number of folds is undefined, it is considered as the K-cross-folds, this technique is also called as repeated holdout method, hence it uses random sampling approach, without replacement the dataset is divided into k random partitions, and each partition consists of approximately n/k number of unique data elements, where n is the total number of data elements and k is the total number of folds. This method has two approaches that are 10-fold cross-validation and leave-one-out cross-validation.

4.3 Bootstrap Sampling

It is a popular way to identify testing and training datasets from the input dataset. It uses the technique of simple random sampling and replacement method, where boot strapping randomly picks data instances from the input dataset, with the possibility of the same data instance to be picked multiple times.

5 Problem Statement

To improve the results the performance of the classification model uses combining classifier methods of different classifier models constructed with committee selection procedure where the solution is to get a robust classifier, a novel ensemble method to be developed to solve the problem. The proposed method works by ensemble of many weak classifiers by the help of boosting algorithms. The results of combined classifiers are studied and show the excellent performance of ensemble based and combining classifier methods. The solution for the problem investigates the overall possibility of using ensemble algorithms to expand the performance of system threat detection and prevention to defend the Deceptive Phishing. The main target is to progress the accurateness and reduce the false-positive rate. To detect the performance measurement, 10-fold cross-validation technique implementation is used for accuracy detection and improvement using data testing mechanisms. Implementing the machine learning algorithms, the goal is to develop a classification and detection of Deceptive Phishing. The problem associated with the classification is noise that interfaces with the reliability with which the features are measured [6].

6 Methodology

Deceptive Phishing is the phishing practice of sending fraudulent communications through emails or malicious URLs, creates fake sites to trap and steal the confidential and sensitive data and installs the malware on victim's machine. It collects the sensitive data using fake websites and these fake websites. The methodology used here is implementing different machine learning and data mining techniques on the dataset for the experimental and evaluation purposes, and we take spam-based dataset. The programming-based data mining and machine learning is done by Java using the WEKA tool.

Machine learning-based approaches are widely experimented in spam-email classification and data mining and identified the phishing websites. Phish tank and spam-base datasets are selected for the study. The random forest algorithm has been preferred for classifying the selected corpuses with features and search methods, and the other classifiers used are K-nearest neighbors and Bayesian classifier algorithms. Data mining technique implemented using the classifications of technical approach revealed talented results. These evaluations are basically based on probability and spam emails and classify the phishing emails and websites.

```
RandomForest

Bagging with 100 iterations and base learner

weka.classifiers.trees.RandomTree -K 0 -M 1.0 -V 0.

Time taken to build model: 2.14 seconds

=== Cross-validation ===
=== Summary ===

Correlation coefficient                   0.917
Mean absolute error                       0.1058
Root mean squared error                   0.1976
Relative absolute error                  22.1403 %
Root relative squared error              40.42   %
Total Number of Instances              4601
```

Screenshot 1: Results of Random Forest on Phish tank

```
RandomForest

Bagging with 100 iterations and base learner

weka.classifiers.trees.RandomTree -K 0 -M 1.0 -V 0.001 -S 1 -do-not-(

Time taken to build model: 1.71 seconds

=== Stratified cross-validation ===
=== Summary ===

Correctly Classified Instances        4394           95.501 %
Incorrectly Classified Instances       207            4.499 %
Kappa statistic                          0.9053
Mean absolute error                      0.1015
Root mean squared error                  0.1947
Relative absolute error                 21.2561 %
Root relative squared error             39.8514 %
Total Number of Instances             4601
```

Screenshot 2: Results of Random Forest on Spam-base Data

To solve the problem statement abovementioned we merged two datasets of the same instances that is 4601. The study examines a spam-email database and the phishing websites database for analysis based on different corpuses that are spambase and Phish tank on various machine learning classifiers that are tested with the same number of instances that is 4601 and combined dataset is known as the deceptive dataset. To obtain the better results the ensemble methods are implemented on the combined deceptive dataset using the boosting, bagging, stacking and voting. Boosting focuses on different algorithms where the system is not performing well, AdaBoost is the most commonly used algorithm implemented to boost the performance, it splits the data into two and implemented on one at first iteration, at next the other part is classified, and later these parts will be clubbed to give result. Fit classifier on data and evaluate overall errors, and error used for calculating weight should be given in final evaluation [7]. The Java code used to start the WEKA is the package weka.api; whereas it also imports some classifiers such as Bayes Naive Bayes, meta.AdaBoostM1, where the different classifiers and boosting methods are used to get more accurate results, such as the code is written to call the csv file to implement the methodology, whereas we can also use arff file for implementation, where the first step is loading the dataset and it may be in the form of csv or arff file format. We use the pubic class for combining the models, where ensemble learners convert the weak learners to the strong learners and allow exception handling techniques along with dataset identification and storage for model implementation. The different pre-processing techniques are used to identify the class, the next step is to code related to boosting, bagging, stacking and voting, and creating a combined classifier is a specific process to identify the goal to get better results. Implementing the combined classifier using Java-WEKA programming and combining the classifiers, then the following output is implemented on the dataset to obtain the better results.

```
=== Summary ===

Correctly Classified Instances       4598              99.9348 %
Incorrectly Classified Instances        3               0.0652 %
Kappa statistic                       0.9986
Mean absolute error                   0.0009
Root mean squared error               0.0181
Relative absolute error               0.1781 %
Root relative squared error           3.6953 %
Total Number of Instances             4601
```

Screenshot 3: The Result Using the Proposed Ensemble Method

7 Mathematical Analysis

In statistics, the mean absolute error is a quantity used to measure how close forecasts or predictions are to the eventual outcomes. The mean absolute error is given by as the name suggests, the mean absolute error is an average of the absolute errors, where it is the prediction and the true value, where A is the actual target and P is the predictive target [8].

$$\frac{\Sigma |A-P|}{n}$$

Root mean square error is given by the following equation

$$\sqrt{\frac{\sum_{i=1}^{n}(A-P)^2}{n}}$$

Relative absolute error is given by the following equation

$$\frac{\sum_{i=1}^{n}(P-a)}{\sum_{i=1}^{n}(\bar{a}-a_1)}$$

where $\bar{a}$ is the mean of the actual value

Relative square error is

$$\frac{\sum_{i=1}^{n}(P-a)^2}{\sum_{i=1}^{n}(\bar{a}-a_1)^2}$$

Root relative squared error is

$$\sqrt{\frac{\sum_{i=1}^{n}(P-a)^2}{\sum_{i=1}^{n}(\bar{a}-a_1)^2}}$$

8 Graphical Analysis

Receiver operating characteristic (ROC) curve is a plot of the true-positive rate against the false-positive rate and characterizes the sensitivity and specificity, plots two parameters that is true-positive rate and characterizes the sensitivity of the same kind of emails that are correctly classified as legitimate emails as the legitimate and wrong classified are false-positive rate. False-positive (FP) rate calculates that the percentage of emails that are legitimate were incorrectly classified by the algorithm as phishing emails. In our implementations, the true-positive rate is that the percentages of the emails of the same kind are correctly classified legitimate emails as the legitimate and the percentages of the phishing emails correctly classified as phishing. Additionally, the false-positive rate is the percentages of the legitimate emails incorrectly classified as phishing emails and the percentages of the phishing emails incorrectly identified as legitimate emails [9].

For the experimental purpose different datasets are taken such as phish tank, spambase and deceptive, to calculate the classification efficiency and performance of the different machine learning classifications that are capable for better performance using the different characteristics such as mean absolute error, root-mean-squared error, relative absolute error and finally root relative absolute error (Table 1).

Table 1 Analysis of errors on different datasets

	Phish tank	Spam-base	Deceptive
Mean absolute error	0.1058	0.1015	0.0009
Root-mean-squared error	0.1976	0.1947	0.0181
Relative absolute error (%)	22.1403	21.2561	0.1781
Root relative absolute error (%)	40.42	39.8514	3.6953

9 Conclusion

Security analyst's all over the world is constantly challenged by the phishing community as new and advanced methods are developing day by day. In this fast-developing environment, it is every researcher's main responsibility to deceive a system that can tackle the situation. In this study, when we compare the different classification algorithms, we have identified the tree-based classifiers as best suitable for the task of phishing URL classification and email classification, and we intend to enhance ability of classifier using different methods such as boosting, bagging, stacking and voting, the system performance further by incorporating a machine learning mechanism. Most of the browsers are built with phishing alert functionality. The classification and filtering majorly based on different types of lists such as the gray list, white list and black list has been an auspicious approach in the past, but dynamic nature of phishing sites demands more efficient methods, which is left for future work.

References

1. F. Ghaffari, H. Gharaee, M.R. Forouzandehdoust, Security considerations and requirements for cloud computing, in *2016 8th International Symposium on Telecommunications (IST)*, Tehran, 2016, pp. 105–110
2. J.V. Chandra, N. Challa, S.K. Pasupuleti, A practical approach to E-mail spam filters to protect data from advanced persistent threat, in *2016 International Conference on Circuit, Power and Computing Technologies (ICCPCT)*, Nagercoil, 2016, pp. 1–5
3. W. Niu, X. Zhang, G. Yang, Z. Ma, Z. Zhuo, Phishing emails detection using CS-SVM, in *2017 IEEE International Symposium on Parallel and Distributed Processing with Applications and 2017 IEEE International Conference on Ubiquitous Computing and Communications (ISPA/IUCC)*, Guangzhou, 2017, pp. 1054–1059
4. J.V. Chandra, N. Challa, S.K. Pasupuleti, Advanced persistent threat defense system using self-destructive mechanism for cloud security, in *2016 IEEE International Conference on Engineering and Technology (ICETECH)*, Coimbatore, 2016, pp. 7–11
5. M.D. Ambedkar, N.S. Ambedkar, R.S. Raw, A comprehensive inspection of cross site scripting attack, in *2016 International Conference on Computing, Communication and Automation (ICCCA)*, Noida, 2016
6. S. Deepika, P. Pandiaraja, Ensuring CIA triad for user data using collaborative filtering mechanism, in *2013 International Conference on Information Communication and Embedded Systems*

(ICICES), Chennai, 2013, pp. 925–928

7. J.V. Chandra, N. Challa, S.K. Pasupuleti, Intelligence based defense system to protect from advanced persistent threat by means of social engineering on social cloud platform. Indian J. Sci. Technol. **8**(28) (2015)
8. N. Abdelhamid, F. Thabtah, H. Abdel-Jaber, Phishing detection: a recent intelligent machine learning comparison based on model content and features, in *2017 IEEE International Conference on Intelligence and Security Informatics (ISI)*, Beijing, 2017, pp. 72–77
9. S. Patil, S. Dhage, A methodical overview on phishing detection along with an organized way to construct an anti-phishing framework, in *2019 5th International Conference on Advanced Computing & Communication Systems (ICACCS)*, Coimbatore, India, 2019, pp. 588–593

A Shape-Based Model with Zone-Wise Hough Transformation for Handwritten Digit Recognition

Dipankar Hazra and Debnath Bhattacharyya

1 Introduction

Character recognition is one type of object recognition and currently an important research area. It can be printed character recognition or handwritten character recognition. Handwritten character recognition is more difficult due to different sizes, directions, etc. Also, character images are written with different widths by different human styles. Same person can write the same character in different ways. Handwritten character recognition is basically of two types such as off-line character recognition and online character recognition. In online handwritten character recognition, a special pen is moving on an electronics surface, and the coordinates from the digits are saved as a function of time. The speed, direction and order of strokes of handwriting are useful information for online handwritten character recognition. This information is not available for off-line character recognition. So, online handwritten character recognition is easier to implement than off-line character recognition. In off-line character recognition, characters are written on a piece of sheet and the sheet is scanned to store the image in binary format. Then different types of object recognition techniques can be followed. Off-line character recognition is also called optical character recognition or OCR. Handwritten digit recognition is dealing with recognition of handwritten digits from 0 to 9. There are numerous successful classification algorithms existing for handwritten digit classification like neural network, SVM, and K-nearest neighbor. All algorithms represent some error due to confusion with digits 1 and 7 or 3, 5 and 8. The challenges of the researchers are to reduce

---

D. Hazra (✉)
Computer Science and Engineering Department, OmDayal Group of Institutions, Uluberia, Howrah, West Bengal 711316, India
e-mail: dipankar1998@rediffmail.com

D. Bhattacharyya
Department of Computer Science and Engineering, Vignan's Institute of Information Technology (A), Visakhapatnam, Andhra Pradesh 530049, India
e-mail: debnathb@gmail.com

© Springer Nature Singapore Pte Ltd. 2020
J. Fiaidhi et al. (eds.), *Smart Technologies in Data Science and Communication*, Lecture Notes in Networks and Systems 105, https://doi.org/10.1007/978-981-15-2407-3_3

this error. Here, shape features computation is done by normalizing signed width vector is successfully introduced. Recognition rate of this method is high due to the use of these extracted shape features. Proposed method recognized digits with lower computation time than other methods because of the preprocessing done involving digit hole. The zone-wise Hough transformation of the digits increases the accuracy of the system.

The paper is arranged as following. In Sect. 2, earlier handwritten digit recognition techniques were surveyed. Section 3 discussed the proposed method in detail. Section 4 displays the result of proposed method and compares it with previous methods. The worth of the proposed method is represented in Sect. 5.

2 Related Works

There are many novel methods for handwritten digit recognition. One such method uses principal component analysis (PCA) for feature extraction and single-layer neural network for classification [1]. MNIST dataset is used for handwritten images. Each pixel of size 28 × 28 is normalized to 0–1. Then the image is converted to greyscale. PCA compresses the number of features for input of neural network. It reduces computation and training time and extracts features accurately. The result shows that neural network with PCA gives the more accurate result than neural network without PCA as classifier. Discrete cosine transformation (DCT)-based approach [2] for capturing discriminating features from handwritten digits also yields good result. Four types of DCT coefficients are used. These coefficients are: upper left corner coefficients, zigzag coefficients, block-based upper left corner coefficients and block-based zigzag coefficients. Features extracted by the principal component analysis are added with the DCT coefficients features to construct the feature vector. Support vector machine is used for classification. Block-based DCT zigzag feature extraction technique yields superior result than other DCT techniques. Handwritten digit and letter recognition using hybrid discrete wavelet transform (DWT) and discrete cosine transform (DCT) feature extraction algorithm and k-nearest neighbor (k-NN) and support vector machine (SVM) classification algorithm achieve high accuracy for digit and letter recognition [3]. MNIST digit dataset and EMNIST letter dataset are used in this experiment. Global thresholding using Otsu's method converts the images in binary form. Isolated pixel removal method is used to remove the noise. Then the image is passed through DWT, and approximation coefficients of DWT are passed through DCT. High value coefficient of DCT is used as feature. SVM and k-NN are two classification techniques applied for classification. SVM has shown better classification accuracy and less computation time than k-NN. Another handwritten recognition system [4] recognizes handwritten numerals using multi-layer feed forward back propagation neural network. Their process consists of three phases: preprocessing, training and recognition. In the preprocessing stage, Gaussian filter is used for removing noise. Then the filtered color image is converted into binary image using OTSU's method. Label numbers are provided to the binary image by

Floyd–Warshall's algorithm. Then binary image is then segmented according to the label number. The segmented binary image is then rescaled using binary interpolation method and fed into neural network architecture for identification. In part-based handwritten digit recognition method, three methods were discussed [5]. The digitized image is considered as a set of parts. Recognition is computed by aggregating recognition of parts. In the first method, Speeded Up Robust Feature (SURF) is used for extract features. The feature-level recognition is done by nearest neighbor (NN). The final digit recognition is done by single majority voting method. In the multiple voting method, different key points of the digits are assigned with different probabilities, and votes in the voting process are different class probabilities. In the class distance method, all the key points of the query digit are in the same class. Kullback–Leibler distance between query digit and reference digit is calculated for digit recognition. The methods show that digit can be recognized without global structure. The methods are also robust against various deformations. The class distance method has the highest recognition rate among the three, whereas multiple voting method is robust than class distance with smaller database size. A Voronoi tessellation-based zoning technique for handwritten digit recognition [6] extracts feature according to optimal zoning distribution found by searching by evolutionary strategy. The zoning divides the image into zones and collects information from each zone. Voronoi tessellation is a zoning technique which partitions image into a set of Voronoi polygons. The gradient direction and pixel density extracted from each zone are used to compute feature for digit recognition. The experiment used SVM classifiers with linear, polynomial and RBF kernels. The digit images are used from MNIST and USPS datasets. It has been shown that Voronoi-based zoning gives better results than regular square zoning for same dataset using the SVM classifier with same kernel. Handwritten digit recognition using orientation and concavity-based features and pooling SVM classifiers [7] yields faster training process and outperforming results. Soft histogram of gradient orientations (SHOG) and histogram oriented gradient (HOG) are two orientation-based descriptors. Both methods use spatial division of image. The 4-connected concavity (4CC) checks north, south, west and east direction of 4-connected neighborhood for constructing feature vector. The 8-connected connectivity (8CC) checks northeast, northwest, southeast, southwest directions of 8-connected neighborhood for constructing feature vector. The 13-bin concavity (13C) follows 4-Freeman directions for constructing feature vector. SVM is used for classification. Multiple classifiers are used and worked on different features, and response of each one is aggregated using pooling process. Pooling process receives scores of each class and aggregates them. Pooling process gives better result than combining the feature extractors of SHOG, HOG, 4CC, 8CC, 13C. The Sum pooling strategy is giving better result than Max or Borda Count. Pooling approach is better for the state-of-the-art CVL database and very competitive for another state-of-the-art database, MNIST database. Some other models of digit recognition also extracted features from numeric digits using HOG features [8, 9]. Then, the digits are classified using support vector machine (SVM). A method uses projection histograms on four directions, i.e., x-axis, y-axis and on lines $y = x$ and $y = -x$ to extract features from handwritten digits [10]. These features were used as input of support vector machine. Bat algorithm and the swarm intelligence algorithm are used for optimal SVM model. This method obtained highly accurate result using simple feature set. A recognition method based on modified chain code histogram-based feature extraction and SVM-based classification [11] gives high accuracy. Chain code histogram by Freeman may be of 4 directions or 8 directions. It gives the directional information from the pixels of the extracted character contour. Directional chain code histogram obtained by differentiating chain code histogram gives the direction changes. Direction turning point is a point where direction changes sharply. Chain code histogram and direction point combine to calculate feature set for handwritten digit recognition. But the chain code suffers the problem of starting point selection on the digit shape. The concavity-based structural features in addition to gradient and chain code features are also used in handwritten digit recognition [12]. The concavity features are calculated by star-like operator that shoot rays in 8 directions and observing hitting point with digit or image boundary. The distance from convex hull to digit boundary is also used as concavity feature. Another way is to find out inner and outer concavity regions by subtracting digit image from convex hull and extract features from each region. Concavity features are complementary features containing limited information to increase accuracy of recognition. SVM or K-nearest neighbor-based classifier used for classification. Another approach focuses on metaheuristics [13] for feature selection. The metaheuristics are applied to MNIST database of handwritten digits. Support vector machine is used for classification task. Reduction of number of features used for digit classification resulted in a small reduction of accuracy rate. Binary Fish School Search achieved better accuracy rate than the other two methods. A method used the nearest neighbor classifier, k-nearest neighbor classifier, k-modified nearest neighbor classifier, fuzzy k-nearest neighbor classifier for digit classification [14]. In k-nearest neighbor algorithm, k neighbors are found. The class with the majority neighbors out of the k-nearest neighbors is the class assigned to the digit. Classification accuracy depends on the value of k. In k-modified nearest neighbor, k neighbors are weighted according to their proximity with the test digit. In fuzzy k-nearest neighbor algorithm, each digit belongs to every class with the membership value depending on the class of its k neighbors. The Hough transformation [15] is useful to detect straight lines in digitized image.

3 System Description

This handwritten digit recognition model is composed of the shape database and functions for training and testing digit images. The system architecture is shown in Fig. 1.

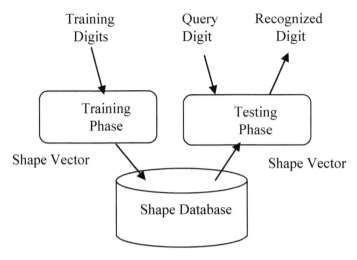

Fig. 1 System architecture for proposed system

3.1 Shape Database

It stores the shape vectors of the training images. Shape vectors are formed by the normalized signed width vector calculation in training phase.

3.2 Training

Image is scanned from top to bottom and left to right. When it gets the first white pixel, extracts the digit using the 8-connected neighbor method and calculates shape vector for storing into digit database. Points on the digits are identified. Then those points are sorted on increasing order of Y, then increasing order of X. For a fixed y coordinate, if left most point X_{left} and right most point X_{right} then the ith width of the digit is $W_i = X_{\text{right}} - X_{\text{left}}$. N widths of the digit from top to bottom in the regular interval are calculated and taken as width vector. Let the width vector be $W = [W_1\ W_2\ \ldots\ W_N]$. Figure 2a, b shows the widths W_1–W_{10} of width vector of sample images 5 and 6, respectively, where $N = 10$. If $W_{\max}$ is the maximum width, then w, the normalized width vector of the digit is following:

$$w = [W_1/W_{\max}, W_2/W_{\max}, \ldots, W_N/W_{\max}] = [w_1, w_2, \ldots, w_N] \quad (1)$$

For increasing accuracy of the proposed method, extra parameter has been added to the normalized width vector. If shape portion is on the completely left side of the central point of the image, negative sign is associated with width vector. Central point is calculated as the average of x-coordinate of leftmost point and rightmost point of

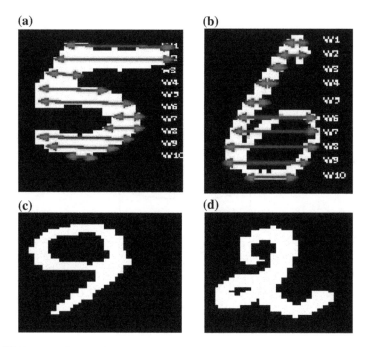

Fig. 2 Width vector for sample images: **a** width vector of 5, **b** width vector of 6, **c** digit 9 with a hole, and **d** digit 2 with a hole with the area less than threshold

the digit. Such width vectors for all the training digits are calculated and stored in the digit database along with digit name.

W_1 and W_8 are the maximum width of digit 5 and digit 6, respectively. After normalization and applying negative sign to some width, width vector for digit 5 becomes

$$w = [1, W_2/W_1, -W_3/W_1, W_4/W_1, W_5/W_1, \ldots, W_{10}/W_1] \qquad (2)$$

After normalization and applying the negative sign to some width, width vector for digit 6 becomes

$$w = [W_1/W_8, W_2/W_8, W_3/W_8, -W_4/W_8, -W_5/W_8, \ldots, W_{10}/W_8] \qquad (3)$$

The shape vectors and the corresponding digit names are stored in the shape database. Two types of recognitions are used in this method, the k-nearest neighbor-based classification and the random forest classification. Nothing to be done in the training stage for a k-nearest neighbor-based classification.

A random forest is a classification method which generates decision trees and a set of rules from subset of given data set. At the time of building decision tree from predefined data classes, each record belongs to a class which is predefined, and one attribute of the given dataset is called class label attribute. The digit name like 'one,' 'two,' etc., are used as the class label attribute of the dataset of the proposed method. A random forest is created for features and steps are as following:

Step 1: Divide dataset into n subsets of data.
Step 2: Select a subset from n subsets, if the subset is not considered before.
Step 3: Randomly select some (m) features from all (M) features, where $m \ll M$.
Step 4: Calculate the best split to find the best splitting attribute A.
Step 5: Split node A to child nodes.
Step 6: Repeat steps 4 and 5, until leaf nodes are reached.
Step 7: Repeat steps 2–6 for n number of times.

3.3 Testing

Testing stage is divided into the presence of holes determination, width vector comparison and zone-wise line characteristics calculation using Hough line transformation.

Holes Presence Determination In a binary image, a hole is represented as the area of black pixels surrounded by white pixels. In the holes, the presence of determination stage and the presence of holes in query digit are detected. If area of the hole is too small, as shown in Fig. 2d, i.e., hole is not part of the digit, but if formed because of handwriting style or any other reason, then that hole is not considered. If hole is present as Fig. 2c, width vector comparison process is carried out for only digits 0, 4, 6, 8, 9; otherwise, width vector comparison process is carried out for 1, 2, 3, 5, 7 and 4 (written without hole). Preprocessing reduces the total testing time and improves classification accuracy.

Width Vector Comparison Width vector of the test digit is calculated following the same process as width vector of training digits. N widths of the digit from top to bottom are calculated and taken as width vector. Let the width vector be $Q = [Q_1\ Q_2\ \ldots\ Q_N]$. Width vector is divided by the maximum width to normalize the width vector. Let the width vector becomes $q = [q_1\ q_2\ \ldots\ q_N]$.

For the k-nearest neighbor-based recognition, the distance between test digit width vector and training digit width vector is calculated using Euclidean distance measure. If d is the distance, then

$$d = \sqrt{(w_1 - q_1)^2 + (w_2 - q_2)^2 + \cdots + (w_N - q_N)^2} \qquad (4)$$

K-nearest neighbors are found out by computing the distances between the test digit and all trained digits. If the majority of these k neighbors are assigned to a digit x, then the digit is recognized as digit x.

For the random forest-based recognition, the class that is detected by maximum number of decision tress will be the class of the digit. Steps for the recognition of the digit by the random forest are as follows:

Step 1: Classify the test feature using decision tree.
Step 2: Store the class of the decision tree.
Step 3: Repeat steps 1 and 2, n times for n decision tree.
Step 4: Calculate votes for each class.
Step 5: The final class is the class which got maximum number of votes.

If one digit is not getting a clear majority, then final recognition is done in zone-wise line characteristics determination stage.

Zone-Wise Line Characteristics Calculation Using Hough Line Transformation The image is divided into different zones, like top, bottom, left, right, top-left, top-right, bottom-left, bottom-right, etc. Horizontal and vertical and other line characteristics of different angles are calculated using Hough line transformation. In Hough transformation, a straight line in (x, y) space is denoted in (r, θ) space as shown in Fig. 3. Here r is perpendicular distance from origin to the straight line and θ is the rotation angle. All points in a line have the same values of (r, θ) parameters. For each point of the image, the parameters are computed for a finite set of angles $\theta = \theta_1, \theta_2, \ldots, \theta_N$. For each angle θ_i, we get radius r_i as follows:

$$r_i = x \cos \theta + y \sin \theta \tag{5}$$

An accumulator matrix was formed where each column represents different values of θ and each row represents different values of r. The highest values in the accumulator matrix represent lines in the image. The horizontal and vertical line characteristics of the digit on different zones are calculated, and the best alternative is recognized as our queried digit. For example, the presence of near vertical lines is more in case of 5 than 3 in the top-left zone of the digit image. Hence, it helps to differentiate between 5 and 3. Similarly, the presence of horizontal and near horizontal

Fig. 3 (r, θ) parameters of a straight line

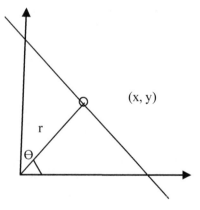

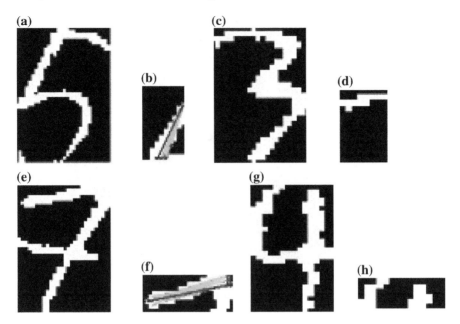

Fig. 4 Hough line characteristics: **a** digit 5, **b** near vertical lines in top-left zone of 5, **c** digit 3, **d** no vertical line in top-left zone of 3, **e** digit 7, **f** near horizontal lines in top zone of 7, **g** digit 4, and **h** no horizontal line in top zone of 4

lines in top zone 7 is more than top zone of 4. Hence, it helps to differentiate between 7 and 4. Figure 4a–d displays Hough line characteristics for vertical and near vertical lines on the top-left zone of sample image of 5 and 3. Figure 4e–h displays Hough line characteristics for horizontal and near horizontal lines on the top zone of the sample images of 7 and 4. Similarly, for different digits, different zones and different types of lines are available; again for some other digits, for some other zones some types of lines are unavailable. This easily helps to distinguish one digit from another digit.

4 Result

Digit images are downloaded from a handwritten digit database [16] by computer vision group at the University of Sao Paulo. Figure 5 shows a sample image consisting of binary handwritten digit 5. For each of the digit, 10 such binary images are there. Seven images for each digit are used for construction of training database, whereas 3 images for each digit are used for testing.

N widths from top to bottom form the width vector, with the negative sign for those which are completely left of the center points. In this way, width vector consists of information of $N * 2$ zones of images and becomes a strong feature vector for

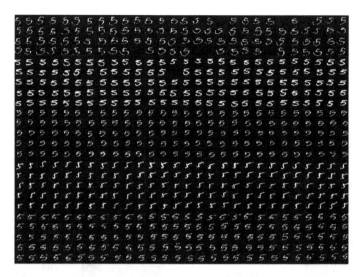

Fig. 5 Sample images of digit 5

classifying digit. The accuracy of this model is 98.3% using k-nearest neighbor-based classification and 98.8% using random forest-based classification, which is comparable to any other state-of-the-art model of handwritten digit recognition in terms of accuracy.

5 Conclusions

The proposed method recognized handwritten digit by means of normalized signed width vector and zone-wise Hough transformation for the feature extraction process. The recognition rate of this method is 98.3% using k-nearest neighbor-based classification and 98.8% using random forest-based classification. The method recognized digits with low computation time because of the hole-based preprocessing. It can be validated against other databases like MNIST database. It encourages researchers for further enhancement of the digit recognition system. Recognition of handwritten character can also be implemented using this method. Again, this method or variation of this method can be useful in handwritten digit recognition in other languages. This method can also be applied to online handwritten digit recognition. Other similar types of shape recognition system can employ the techniques used in proposed system.

References

1. V. Singh, S.P. Lal, Digit recognition using single layer neural network with principal component analysis, in *IEEE Asia-Pacific World Congress on Computer Science and Engineering*, Fiji, 2014, pp. 1–7
2. B. Qacimy, M. Kerroum, A. Hammouch, Handwritten digit recognition based on DCT features and SVM classifier, in *Proceedings of World Conference on Complex Systems*, Agadir, Morocco, 2014, pp. 13–16
3. P. Ghadekar, S. Ingole, D. Sonone, Handwritten digit and letter recognition using hybrid DWT-DCT with KNN and SVM classifier, in *IEEE International Conference on Computing Communication Control and Automation (ICCUBEA)*, Pune, India, 2018, pp. 1–6
4. P.P. Selvi, T. Meyyappan, Recognizing handwritten numerals using multilayer feed forward back propagation neural network. Int. J. Comput. Technol. Appl. **3**(6), 1939–1944 (2012)
5. S. Wang, S. Uchida, M. Liwicki, Y. Feng, Part based method for handwritten digit recognition. Front. Comput. Sci. **7**, 514–524 (2013). Springer, Heidelberg
6. S. Impedovo, F.M. Mangini, G. Pirlo, A new adaptive zoning technique for handwritten digit recognition, in *ICIAP 2013, Part I*, ed. by A. Petrosino. LNCS, vol. 8156 (Springer, Heidelberg, 2013), pp. 91–100
7. J.M. Saavedra, Handwritten digit recognition based on pooling SVM-classifiers using orientation and concavity based features, in *CIARP 2014*, ed. by E. Bayro-Corrochano, E. Hancock. LNCS, vol. 8827 (Springer, Cham, 2014), pp. 658–665
8. R. Ebrahimzadeh, M. Jampour, Efficient handwritten digit recognition based on histogram of oriented gradients and SVM. Int. J. Comput. Appl. **104**(9), 10–13 (2014)
9. K. Banjare, S. Massey, Numeric digit classification using HOG feature space and multiclass support vector machine classifier. Int. J. Sci. Res. Educ. **4**(5), 5339–5345 (2016)
10. E. Tuba, M. Tuba, D. Simian, Hand written digit recognition by support vector machine optimized by bat algorithm, in *Proceedings of the 24th International Conference on Central Europe on Computer Graphics, Visualization and Computer Vision*, Plzen, Czech Republic, 2016, pp. 369–376
11. Q. You, X. Wang, H. Zhang, Z. Sun, J. Liu, Recognition method for handwritten digits based on improved chain code histogram feature, in *Proceedings of the 3rd International Conference on Multimedia Technology*, Guangzhou, China, 2013, pp. 438–445
12. M. Karic, G. Martinovic, Improving offline handwritten digit recognition using concavity based features. Int. J. Comput. Commun. **2**(8), 220–234 (2013)
13. L. Seijas, R. Carneiro, C. Santana, L. Soares, S. Bezzera, C. Bastes-Filho, Metaheuristics for feature selection in handwritten digit recognition, in *Latin America Congress on Computational Intelligence*, Curitibia, Brazil, 2015, pp. 1–6
14. V.S. Devi, M.N. Murty, *Pattern Recognition: An Introduction* (University Press, Hyderabad, India, 2011)
15. R.O. Duda, P.E. Hart, Use of Hough transformation to detect lines and curves in pictures. Commun. ACM **15**, 11–15 (1972)
16. Computer Vision Group, *Handwritten Digit Database* (University of Sao Paulo, 2005)

Deducted Sentiment Analysis for Sarcastic Reviews Using LSTM Networks

Labala Sarathchandra Kumar and Uppuluri Chaitanya

1 Introduction

Artificial intelligence is pacing forward in numerous fields like e-commerce, medicine, technological advancements, etc. For any e-commerce empire, their customer satisfaction is their primary motive. To scale the amount of success, their product has achieved, knowing to what extent a customer is satisfied with their product being necessary. So, all the e-commerce Web sites get to know the customer opinions through ratings, reviews and suggestions. What makes it challenging is not the collection of opinions, but the analysis of the large number of opinions collected across the globe [1]. This brings sentiment analysis into picture that analyzes the opinions and predicts the sentiment: whether that review is either positive or negative toward the product or service.

Sentiment analysis is one of the natural language processing fields that can build systems which attempt to extract and determine opinions at intervals text. Currently, sentiment analysis may be a concept of nice interest and precipitating development due to its several sensible applications. With the assistance of sentiment analysis systems, a vast collection of unstructured information might be mechanically reworked into structured knowledge of public point of views [2]. Sarcasm can be described as a mocking or a criticizing opinion portrayed in a humorous fashion. Sarcasms appear as they contradict their actual essence [3]. Though sarcasm is of negative sentiment, they are embellished as if they possess a positive sentiment. In case of sarcasm, the actual opinion the sarcasm conveys can only be understood by humans. When people opine reviews with such sarcasms involved, then they are called **sarcastic reviews**. Sentiment analysis can predict the sentiment of ordinary reviews accurately. But when it comes to sarcastic reviews, the reviews get twisted that the traditional algorithms used by the sentiment analysis may not be able to predict

---

L. Sarathchandra Kumar (✉) · U. Chaitanya
Department of Computer Science and Engineering, Vignan's Institute of Information Technology, Visakhapatnam, Andhra Pradesh, India
e-mail: sarathlabala888@gmail.com

the correct sentiment [4]. Taking examples like "This shirt suits you just as it suits your grandfather", "This movie is so interesting that you can buy a pillow instead of popcorn" are clearly sarcastic that they cannot be analyzed by the sentiment analysis as sarcastic ones. These reviews are considered like other normal reviews, and sentiment is predicted inaccurately [5].

2 Related Work

2.1 Deep Learning

Deep learning makes computers learn by themselves what human mind is good at doing inherently: learning by examples. Deep learning, a machine learning technique, is prevalent everywhere from autonomous cars to handy consumer devices, e.g., phones, tablets and hands-free devices. Deep learning enables autonomous cars in recognizing roads and distinguishing objects in their path. Deep learning is obtaining lucrative attention these days. It is achieving the results that once were next to impossible. A deep learning computer model learns by itself on how to perform classification-related tasks with sources like text, voice and images. Deep learning models are well known to accomplish remarkable precision, in some cases unquestionably more exact than human-equivalent performance. Models from deep learning are built with the artificial neural network's architecture that contain various layers and are trained with large sets of labeled data [1].

Though deep learning first came into existence theoretically in 1943, there are two main considerations why it is becoming useful in recent days: Deep learning requires mammoth amounts of **labeled data**. For example, development of autonomous car requires a greater number of pictures of roads, traffic, pedestrians and rules and more hours of videos regarding them. Deep learning demands massive **computing capability** and high-performance GPUs with a proper parallel architecture, especially suitable for deep learning. Combination of grids, clusters or even cloud computing can provide considerable computational power, development efforts and time required for training a deep neural network reduces from days to hours or even less [6]. Most of the deep learning methods use artificial neural network's architectures; because of this, deep learning models are often mentioned as deep neural networks.

The word "deep" usually indicates a number of hidden (inner) layers in a neural network. Conventional neural networks contained only two to three hidden layers, while the deep networks can have as many hidden layers as required as shown in Fig. 1.

Deep learning models, as said earlier, learn knowledge from features of the data directly without any need for extracting features manually. The conventional algorithms that are used in sentiment analysis were machine learning algorithms. Machine learning algorithms emphasize on algorithms, and their working rather than giving

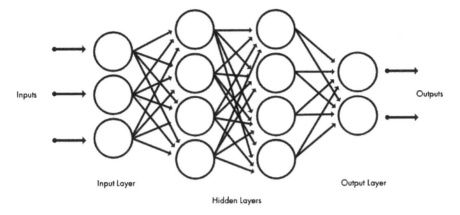

Fig. 1 Deep neural network

importance to the data that is involved. So it is clear that machine learning algorithms are less efficient at effectively handling the sarcastic reviews. But deep learning focuses purely on data and how the data needs to be handled. We are using LSTM networks (Fig. 2), a neural network from deep learning [6].

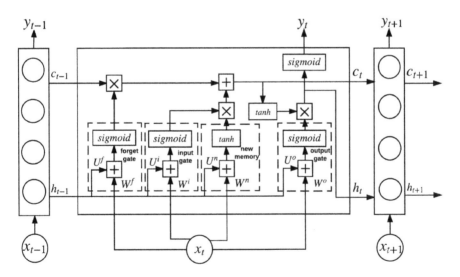

Fig. 2 LSTM neuron

2.2 LSTM Layers

As said earlier, sarcastic reviews cannot be analyzed by the sentiment analysis properly. So the model needs to be trained by a number of sarcastic reviews to familiarize them to the model. A sarcastic review can be determined only after the entire review is remembered till the end and analyzed. But the conventional sentiment analysis cannot store long sequences of reviews. This results in prediction of some sentiment even before the entire review is read by the model. LSTM layers can solve this problem.

In standard repeated neural network, throughout the gradient back-propagation, as the range of steps increase, the gradient signal gets multiplied. Therefore, the gradient signals will end up being increased numerous times with the weight matrix, which supplies the connections form neuron to neuron of the hidden layer in an exceedingly recurrent network. This says the value magnitude of weights within this transition matrix will powerfully impact the method of learning [7].

If the weights within the weight matrix are too little, it can cause vanishing gradient, where as the gradient signal gets too little that typically learning will become terribly slow. Sometimes, it may completely stop working. As a result, the phase of learning dependencies in long term from the data becomes even more difficult. Similarly, if the weights of the matrix are too large, it can cause the gradient signal to diverge from the learning process. This is often called as *exploding gradient*.

These questions are answered by the LSTM model by introducing a brand new structure known as memory cell. A memory cell (Fig. 3) is formed of three gates, i.e., an input gate, output gate, forget gate and, in addition, a neuron with a self-recurrent connection. The self-recurrent connection has 1.0 as weight. It maintains the state of a memory cell, remains constant from one time step to another. The first gate, input gate, deals with incoming signal and checks whether to change the memory cell state or blocks the same. The next gate, output gate, deals with whether to permit the state of the memory cell to own an impact on other neurons or forestall it. The final gate, the forget gate, will enable the cell to recollect or forget the previous state, as per need.

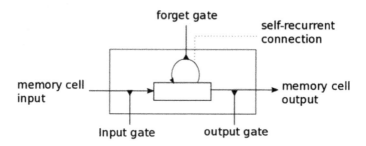

Fig. 3 Memory cell

The beneath conditions depict how memory cells of a layer are updated at each time step "t":

- x_t is given to the memory cell layer at time t
- $W_i, W_f, W_c, W_o, U_i U_f, U_c, U_o$ and V_o are weight matrices
- b_i, b_f, b_c and b_o are bias vectors.

Firstly, the input gate value i_t and $\tilde{C}_t$ the candidate value for the states of the memory cells at given time t are computed:

$$i_t = \sigma(W_i x_t + U_i h_{t-1} + b_i)$$
$$\tilde{C}_t = \tanh(W_c x_t + U_c h_{t-1} + b_c)$$

Next the activation function f_t of the forget gates of memory cell at given time t can be computed:

$$f_t = \sigma(W_f x_t + U_f h_{t-1} + b_f)$$

As the values of the activation function i_t of input gate, the activation f_t of forget gate and the value of candidate state $\tilde{C}_t$ are computed, the new state of a memory cell at given time t can be:

$$C_t = i_t * \tilde{C}_t + f_t * C_{t-1}$$

The values of output gates and respective outputs are computed with the help of new state obtained, as follows:

$$o_t = \sigma(W_o x_t + U_o h_{t-1} + V_o C_t + b_o)$$
$$h_t = o_t * \tanh(C_t)$$

2.3 Advantages of Proposed System

It is estimated that most of the world's data are unstructured and unorganized. These data comprise of tweets, reviews, emails and chats [8]. These texts are so large that they are tough, long and pricey to analyze, understand and sort through.

Sentiment analysis systems permit such unstructured matter knowledge to form sense and help automating the business processes, obtaining unjust insights and then saving time of manual data processing, by producing the most effective results.

Some more advantages of sentiment analysis using LSTM network:

A. *Scalability*

Sorting and processing hundreds of thousands of reviews, opinions and tweets manually are the most hectic task. Sentiment analysis with LSTM allows processing of enormous data in a productive and cost-effective way.

B. *Realtime Analysis*

Sentiment analysis may be accustomed to determine vital data that enables situational awareness during specific situations in real time. How many people have a positive opinion on a movie? How satisfied are the thousands of customers with a product? What is the current trending topic on social media? A sentiment analysis system with LSTM can help not only in immediately identifying these kinds of situations and taking actions but also can memorize large volumes of data at once and act accordingly.

C. *Consistent Criteria*

Criteria of evaluating a sentiment of a chunk of text vary from person to person. Analyzing a sentience could be highly influenced by personal experiences, prejudiced thoughts and partisan beliefs. However, by employing a centralized sentiment analysis system, one can apply similar criteria to the whole knowledge. This helps to scale back errors and then can improve data consistency.

LSTM is a deep learning neural network which has a cell state that makes it stand apart from rest of the neural network algorithms which solves the key problem of long-term dependency remembering.

The initial phase of our LSTM network is to choose the data that we are going to discard from the cell state. It is done by the sigmoid layer called the "forget gate." It verifies h_{t-1} and x_t, yields a number in the range from 0 to 1 for every member in the cell state C_{t-1}. A 1 represents "completely keep this," while a 0 represents "completely eliminate this."

Now consider the case of a language model attempting to foresee the following word dependent on all the past ones. In such a case, the cell state would possibly embody the gender of this subject, so that the proper pronouns will be used. Once a new subject is seen, the gender of the previous subject can be forgotten.

The subsequent stage is to choose what new data will be kept in the cell state. This consists of two parts. Initially, a sigmoid layer known as the "input gate" chooses which esteems are to be updated. Secondly, a vector of new candidate values is made by tanh layer, "~C_t." In the following step, these two are consolidated to make an update to the state.

3 Problem Deduction

In order to obtain sentiments, we remove unrelated data and data which do not help in right sentiment prediction. We consider the text reviews and tokenize them into the words. We are creating sequences of individual words.

As stated, conventional sentiment analysis does not accept text reviews as sequence problems. So by using NLP (Natural Language Processing), we are converting text reviews in human understandable language into a vector [9, 10]. Thus, we are considering the reviews as sequential problems. These sequences are inputted to the LSTM layers to find the patterns that can be recognized through training. Afterward, LSTM provides us the solution in two possible ways:

a. Boolean solution: Predicting whether the given sentiment is positive or negative that is 0 or 1 depending upon the sentiment it conveys.
b. Probabilistic solution: Entire review is predicted based on probability. If the review has positive sentiment, then the probability of positive sentiment is high, else negative sentiment will have a high probability.

4 Methodology

a. **Dataset Collection**

We are considering AMAZON REVIEW dataset. We are using an open sarcastic review collection called "The Sarcasm Corpus" for this experiment. The corpus consists of hundreds of reviews on products from the e-commerce website Amazon, and the reviews are labeled as ironic and regular. We can consider the ironic reviews as sarcastic reviews in this context. Each review contains fields such as ratings, title, product and their view [11].

Initially, we start out with making a comma-separated values (CSV) file of Amazon review dataset. Next we assign sentiments to all the reviews regarding the gist they possess.

b. **Data Preprocessing**

The first step manages basic cleaning operations, which consists of removing unnecessary or unrelated words and sending them for the next stage of analysis. All the operations of this module are executed to try to make the text uniform. It is called "basic cleaning."

Next we put word variations like "smart," "smartly," "smartest" and "smarter" all into one place, effectively increasing the relevance of the concept of "smart" by decreasing entropy. In other words, we consider parts of speech like nouns, verbs and adverbs which have the equal radix in the same way. This is called "stemming."

Upon preprocessing unnecessary words like pronouns, articles are refined out, which are known as "stop words." The presence of these words only leads to reduction in accuracy [9].

c. **Tokenization**

Tokenization is that the method of breaking down a given text or string into items like words, keywords, phrases, symbols and alternative parts referred to as tokens. Tokenization is that the method of changing text as tokens before remodeling it into vectors. It additionally becomes easier to separate excess tokens [12].

d. **Embedding LSTM Layers**

Long Short-Term Memory (LSTM) network is an extended version of RNN and can learn long-run dependencies. For an LSTM, along with input vectors hidden states are also fed from previous neurons to the following neurons. What separates LSTM from RNNs (Recurrent Neural Networks) is LSTM neuron that has three gates: (1) forget gate, (2) input gate and (3) output gate. The forget gate discards words that are not needed for sentiment prediction. The input gate accepts the input vector of words. The output gate outputs the sentiment. In addition to these gates, LSTM has cell state which can remember long-run dependencies. In LSTM, hidden state and cell state change from a previous state to a present state along with the input vectors. The presence of these cell states makes LSTM stand out of other neural networks. This makes LSTM capable of remembering long sequences, while the other networks fail to do so. This is how the solution for obtaining sentiment for sarcastic reviews is procured [13].

5 Results and Analysis

The model built has to be trained with a large number of reviews to familiarize the model with the sarcastic reviews and their analysis. The accuracy achievement depends upon factors like the number of epochs, maximum number of words considered and number of reviews. Some of the test cases are shown in Table 1.

Figure 4 shows that the maximum accuracy obtained with 8 lakh reviews considered is 84.4%.

Table 1 Accuracy achievements for different test cases

No. of epochs	Max words	No. of reviews	Training accuracy	Test accuracy
10	20	500,000	**85.13**	**76.95**
20	20	800,000	**88.55**	**76.66**
10	100	500,000	**89.47**	**84.29**
20	100	800,000	**92.34**	**84.4**

Numerics in bold are the accuracies in increasing order

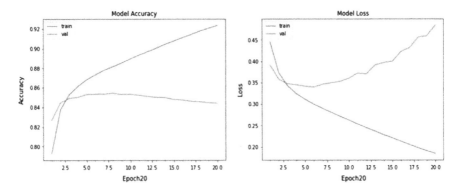

```
Accuracy:     84.4%

              precision  recall  f1-score  support

           0    0.88      0.86     0.87     95534
           1    0.80      0.82     0.81     64466

   micro avg    0.84      0.84     0.84    160000
   macro avg    0.84      0.84     0.84    160000
weighted avg    0.84      0.84     0.84    160000
```

Fig. 4 Evaluation metrics

Fig. 5 Accuracy and loss graphs

We listed all the parameters of this optimal result and the fitting graphs (Fig. 5) of this case.

6 Conclusion and Future Scope

We observe how difficult it is to differentiate a sarcastic sentence from a regular sentence given the similarity of the sentence structure. We also understand the importance of knowing the context in which a sentence is used in order to categorize a sentence as a sarcastic sentence or a regular sentence.

We also infer that conventional feature such as sentiment of a sentence, parts of speech tags, word unigrams and bigrams and punctuation marks are helpful, but not enough to recognize sarcasm in text.

From the above given test cases, it can be observed that by considering 8 lakh reviews, an accuracy of 84.4% is obtained. If a number of reviews to be trained are increased further, accuracy may be improved beyond 84.4%.

6.1 Future Scope

The system we designed has the following future enhancements. It can be used to analyze the sentiments on emoji/smiley in near future which will determine whether the given review is positive or not. To determine the neutrality of text which is a tedious task, prospective improvement can be made to our data collection and methods of analysis. In future, research can be done with more possible improvements such as refined data and far more accurate algorithms.

References

1. L. Zhang, S. Wang, B. Liu, Deep learning for sentiment analysis: a survey
2. B. Liu, *Sentiment Analysis: Mining Opinions, Sentiments, and Emotions* (The Cambridge University Press, 2015)
3. S. Jain, A. Ranjan, D. Baviskar, Sarcasm detection in amazon product reviews
4. P. Tungthamthiti, K. Shirai, M. Mohd, Recognition of sarcasms in tweets based on concept level sentiment analysis and supervised learning approaches
5. L. Peled, R. Reichart, Sarcasm SIGN: interpreting sarcasm with sentiment based monolingual machine translation, in *Proceedings of the Annual Meeting of the Association for Computational Linguistics (ACL 2017)*, 2017
6. I. Goodfellow, Y. Bengio, A. Courville, *Deep Learning* (The MIT Press, 2016)
7. Z. Wang, Y. Zhang, Opinion recommendation using a neural model, in *Proceedings of the Conference on Empirical Methods on Natural Language Processing (EMNLP 2017)*, 2017
8. L. Deng, J. Wiebe, Recognizing opinion sources based on a new categorization of opinion types, in *Proceedings of the International Joint Conference on Artificial Intelligence (IJCAI 2016)*, 2016
9. L. Gui, R. Xu, Y. He, Q. Lu, Z. Wei, Intersubjectivity and sentiment: from language to knowledge, in *Proceedings of the International Joint Conference on Artificial Intelligence (IJCAI 2016)*, 2016
10. H. Rashkin, E. Bell, Y. Choi, S. Volkova, Multilingual connotation frames: a case study on social media for targeted sentiment analysis and forecast, in *Proceedings of the Annual Meeting of the Association for Computational Linguistics (ACL 2017)*, 2017
11. M. Chen, Y. Sun, Sentimental analysis with amazon review data
12. B. Pang, L. Lee, Opinion mining and sentiment analysis. Found. Trends Inf. Retr. (2008)
13. L.C. Yu, J. Wang, K.R. Lai, X. Zhang, Refining word embeddings for sentiment analysis, in *Proceedings of the Conference on Empirical Methods on Natural Language Processing (EMNLP 2017)*, 2017
14. P. Nakov, A. Ritter, S. Rosenthal, F. Sebastiani, V. Stoyanov, Sentiment analysis in Twitter

Automatic Identification of Colloid Cyst in Brain Through MRI/CT Scan Images

D. Lavanaya, N. Thirupathi Rao, Debnath Bhattacharyya and Ming Chen

1 Introduction

A tumor that contains gelatinous substances in the brain is a colloid cyst. In memory, the problems like diplopia and cognitive disturbances, extreme instances, sudden mortality are the serious issues caused due to the presence of cyst in brain. Colloid cyst constitutes 0.5–1% of intracranial tumors. Symptoms of intermittent symptoms are typical for this lesion. Untreated cyst stress could lead to hernia of the brain. The anterior aspect of the third ventricle, which originates from the top of the ventricle, is almost always found behind the foramen of Monro. It may trigger blocking hydrocephalus and increased intracranial pressure due to its place. The symptoms may include headache, vertigo, and fading.

In biomedical field, the identification of any disease in human at fraction of seconds is very important to save the human life at early stage. The image processing plays a good role in biomedical field in the identification of any disease. The image processing can be defined as performing some operations in the image to analyze the information or to identify the information by techniques involved in image processing [1]. The images are produced from MRI/CT scan devices. Based on the disease occurred in the human and area of identification, the doctors suggest the patient

---

D. Lavanaya (✉) · N. T. Rao · D. Bhattacharyya
Department of Computer Science and Engineering, Vignan's Institute of Information and Technology (A), Visakhapatnam, AP 530049, India
e-mail: d.lavanya094@gmail.com

N. T. Rao
e-mail: nakkathiru@gmail.com

D. Bhattacharyya
e-mail: debnathb@gmail.com

M. Chen
Harbin University of Commerce, Harbin, China
e-mail: mingchen@163.com

to take either MRI or CT scan. Considering these images, the image processing techniques are applied to retrieve the information from the image [2].

Image processing is a way to conduct certain picture operation so that an improved picture can be obtained or some helpful data can be extracted. It is a signal processing form, in which images are input and the output can be images or features [3, 4]. The processing of images today is one of the fast-growing techniques [5]. It forms a key area of studies in the areas of engineering and IT. Two techniques, namely analog and digital imaging, are used for image processing. For difficult copies such as printouts and pictures, analog picture processing can be used [6].

2 Literature Survey

In the past, various algorithms have been developed to automate the system of detecting cysts and tumors in the brain by using image processing techniques like threshold segmentation, edge detection, clustering-based segmentation, watershed segmentation, and many more.

Karishma Sheikh, Vidya Sutar, and SilkeshaThigale in their paper proposed a system that used a pixel-to-pixel comparison, the grayscale, and K-means segmentation algorithm to detect tumor from MRI images [7–9]. They used clustering to differentiate between affected and unaffected cells.

Mr. Lalit P. Bhaiya, Ms. SuchitaGoswami, and Mr. VivekPali classified abnormalities in magnetic resonance brain images by developing a hybrid model that combined advantages of both artificial neural networks and fuzzy logic [10]. In this system, textural features were extracted using the principle.

Debapriya Hazra et al. [11] had considered the MRI/CT scan images, and by using the MATLAB programming and image processing models, they tried to identify the presence of colloid cyst in brain images.

D. Lavanya, N. Thirupathi Rao, and Debnath Bhattacharyya have implemented a system to identify the presence/absence of cyst using different approaches, i.e., by matrix generation from monochrome image by using image processing techniques and Java programming [12].

2.1 Observations from Previous Works

From the various works mentioned above, the matrix method is no longer used by the authors to combine monochrome pictures to identify the cyst presence with MRI and CT scan images in the human brain. We have therefore attempted to take into account these observations, and a new monochrome picture matrix method with a new algorithm has been used to determine whether cysts are found in human brain MRI/CT pictures.

3 Proposed Work

In the current work, the new algorithm that identifies the colloid cyst in the brain of all age groups has been considered and the algorithm can be observed as follows in order to identify the cyst in the mind using MRI/CT scan images.

3.1 Algorithm

i. Start.
ii. Read input MRI/CT scan image.
iii. Grayscale conversion.
iv. Monochrome conversion.
v. Matrix generation.

3.2 Study of Algorithm

- Firstly, in acquisition stage the MRI/CT scan images are taken to identify the cyst in brain. This would be considered as an input image.
- In the second stage, i.e., preprocessing stage, the input image is converted into grayscale by using "average method."
- At the time of segmentation phase, the grayscale image is considered as an input and it is changed into monochrome image by "Otsu threshold method."
- To identify the cyst in MRI/CT scan images more clearly, the matrix is generated from the monochrome image obtained at segmentation phase.
- The matrix contains binary values, i.e., zeros and ones. The ones indicate the presence of the cyst, and the zeros indicate absence of colloid cyst.

3.3 Flowchart

The flowchart helps us to understand in a better way about the process in the identification of colloid cyst even for the people who are not aware of image processing (Fig. 1).

Fig. 1 Flowchart

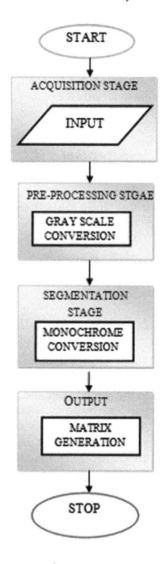

4 Results

In this paper, we have taken two images that containing cyst and not containing the cyst in testing the system that is capable of identifying the colloid cyst in brain using MRI/CT scan images of all age groups and produce accurate result in less time. The results can be viewed in Figs. 2 and 3.

Case Study 1: Identifying the cyst in input image

It is clear that the input image considered for testing the system produces correct output. The colloid cyst always occurs at the center of the third ventricle in brain.

Fig. 2 Input image

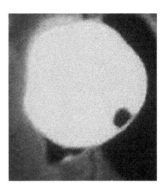

Fig. 3 Grayscale picture

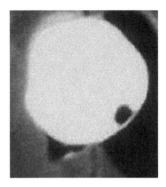

So therefore we can conclude that the ones appear only at the center. In the above matrix, i.e., Fig. 4, the ones appeared irrespective of cyst size as there is giant cyst observed in Fig. 5.

Case Study 2: Identifying the cyst in color MRI image

Fig. 4 Monochrome picture

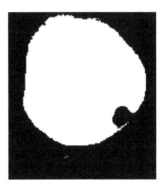

Fig. 5 Matrix generation

Fig. 6 Input color image

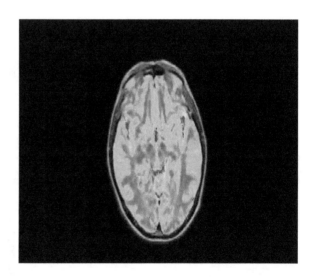

The system is capable of identifying the cyst in color MRI image. We can see the matrix produced contains only zeros. This says the input mage does not contain colloid cyst (Figs. 6, 7, 8, and 9).

Fig. 7 Grayscale picture

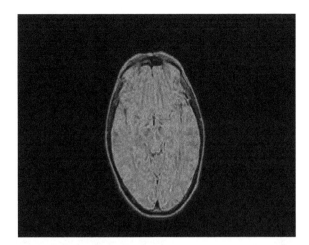

Fig. 8 Monochrome picture

5 Conclusion

The proposed system is capable of identifying cyst in all age groups automatically in less time by using image processing techniques. The cyst occurs in the brain at third ventricle, i.e., center of the brain. Considering this point, the matrix generated from monochrome image will contain ones when cyst is present in the input image. Those ones are located in the center of the matrix irrespective of cyst shape in input image. Automatic detection of cyst using "Otsu threshold method" results in less time compared to "fixed threshold method."

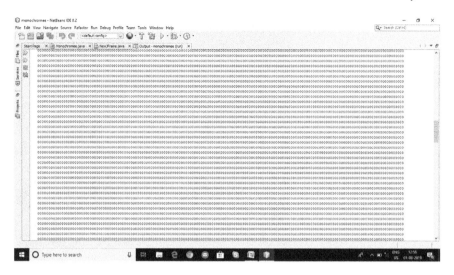

Fig. 9 Matrix generation

References

1. E.E.M. Azhari, M.M. Hatta, Z.Z. Htike, S.L. Win, Brain tumor convergence and services. IJITCS **4**(1), 1–11 (2014)
2. http://www.ncbi.nlm.nih.gov/pmc/articles/PMC1550234/. Accessed 03 July 2018
3. https://en.wikipedia.org/wiki/Image_noise#Low_and_high-ISO_noiseexamples. Accessed 05 July 2018
4. http://www.mecs-press.org/ijisa/ijisa-v5-n11/IJISA-V5-N11-3.pdf. Accessed 07 July 2018
5. V. Kshirsagar, J. Panchal, Segmentation of brain tumor and its area calculation. Int. J. Adv. Res. Comput. Sci. Softw. Eng. **4**(5), 528–529 (2014)
6. https://en.wikipedia.org/wiki/Cyst. Accessed 16 July 2018
7. Q. Javed, A. Dutta, Third ventricular colloid cyst and organic hypomania, in *Progress in Neurology and Psychiatry*, (2014), p. 18
8. http://www.medicalnewstoday.com/articles/181727.php. Accessed 17 June 2018
9. http://www.abta.org/secure/resource-one-sheets/cysts.pdf. Accessed 18 June 2018
10. http://mstrzel.eletel.p.lodz.pl/mstrzel/pattern_rec/filtering.pdf. Accessed 06 July 2018
11. D. Hazra, D. Bhattacharyya, H.J. Kim, Detection of colloid cyst in brain through image processing techniques. Int. J. Multimed. Ubiquitous Eng. **11**(9), 343–354 (2016)
12. D. Lavanya, N.T. Rao, D. Bhattacharyya, Generalized detection of colloid cyst in brain using MRI/CT scan images. Int. J. Innov. Technol. Explor. Eng. (IJITEE) **8**(4) (2019). ISSN: 2278-3075

A Detailed Review on Big Data Analytics

Eswar Patnala, Rednam S. S. Jyothi, K. Asish Vardhan
and N. Thirupathi Rao

1 Introduction

Big data is a concept used to deal with huge volumes of data. The data comprises of both structured and unstructured data, and these are difficult to process by using tradition databases and software method [1].

Big data was introduced by John Graunt in 1663, when he was doing research on bubonic plague, a viral diseases which spreaded all over the Europe. He performed some statistical methods on the data gathered regarding the disease and stood as the first person, who performed statistical analysis. Later in 1800, this statistics has been expanded by collecting and analyzing data. The evolution of modern technologies made big data as one of the emerging technologies of computer science. Most of the devices like smart phones, computers, and sensors made the usage of the big data in a great deal. As these devices produce enormous amount of data, to handle such huge data, we need big data. One of the major contributions in the evolution of big data is social media [2].

E. Patnala (✉)
Department of Information Technology, UCEV-JNTUK, VZM, India
e-mail: eswar.patnala@gmail.com

R. S. S. Jyothi · K. Asish Vardhan
Department of Computer Science and Systems Engineering, College of Engineering (A), Andhra University, Visakhapatnam, India
e-mail: sujanjyohi26@gmail.com

K. Asish Vardhan
e-mail: ashi.mintu@gmail.com

N. Thirupathi Rao
Department of Computer Science and Engineering, Vignan's Institute of Information Technology (A), Visakhapatnam, AP, India
e-mail: nakkathiru@gmail.com

1.1 Foundation of Big Data

The major reason for foundation of big data is information or data. In 1880, a census bureau in USA faced a problem regarding data. They have come to conclusion that, it would take more than 10 years to handle and process the data which was generated from 1880 census. So, to solve this problem, a man who was the employee of the census bureau developed a machine called Hollerith tabulating machine. By using this machine, the processing time has been reduced from 10 years to 3 months.

Later in 1927, an Austrian–German engineer named Fritz Pfleumer introduced magnetic tape technology for storing the information and he got patents for this technology in 1928. During 1943, British scientists introduced a machine called colossus to intercept the messages from Germans during World War II. So, this machine reduces the workload from weeks to hours and this colossus which is a data processor used for data handling purpose [3].

Big data can be represented in three forms, they are

(i) Structured data
(ii) Unstructured data
(iii) Semi-structured data.

i. **Structured data**: If a data has been stored, accessed, and processed in a specified format, then it is called as structured data. The data stored in relational database is perfect example for structured data [2] which can be observed in Table 1.
ii. **Unstructured data**: Data which was not in specified form or structure is denoted as unstructured data. Generally, unstructured data is a combination of text, image, audio, and video files, etc. We can simply represent as heterogeneous data [2].
iii. **Semi-structured data**: Semi-structured data is the combination of both structured and unstructured data. Generally, semi-structured data represented in structured data form but it is not actually defined [2].

2 Characteristics of Big Data

The characteristics of big data are represented by 4Vs and they are [2].

i. Volume
ii. Variety

Table 1 Employee schema

Emp_ID	Emp_Name	Gender	Dept	Salary_In_Lakhs
1	ABC	M	Production	1,00,000
2	XYZ	F	HR	1,43,000
3	EFG	M	Finance	2,10,000

iii. Velocity
iv. Variability.

i. **Volume**: It is one of the basic characteristics when we are dealing with big data. When processing a query, data size plays an important role in finding the value out of the data [2].
ii. **Variety**: The next crucial aspect of big data is the variety. Here, it comprises of both structured and unstructured data. During initial days, spreadsheets and databases are only sources of data for most of the application. But nowadays, we have variety of data like images, audio, videos, etc., and when dealing with unstructured data, we shall apply some mining and storage methods for analyzing the data.
iii. **Velocity**: It is one of the important characteristics of big data, as it is dealt with speed of processing the data in order to meet the deadlines or to display the results [2].
iv. **Veracity**: It deals with the accuracy of data, when it trust worthy or not. As data is generated from numerous applications, we cannot judge them as trust worthy. This characteristic removes all abnormalities, inconsistencies, and biases in the data [4].

3 Benefits of Big Data

The following are some of the benefits of big data

(i) Business can utilize outside intelligence while taking decisions
(ii) Improved customer service
(iii) Early identification of risks to the product/services if any
(iv) Better operational frequency.

4 Modules of Hadoop

Hadoop Distributed File System (HDFS)

One of the main features of Hadoop is distributed file system, and it is called as HDFS. By using HDFS, data is distributed over all machines. So that, if there is loss of data, we can recover from other machines and also by distributing the data there will be the high availability of data for the applications which are running in parallel [5].

Where to use HDFS

(1) Very large files: HDFS method is applied to the files which are of large size like MBs and GBs [5].
(2) Streaming data access: HDFS method works in the pattern of write once and read many times [5].
(3) Commodity hardware: HDFS method works on low-cost devices also [5].

Where not use HDFS

(1) **Low Latency data access**: HDFS method should not use for the applications which take less time to access, because this method is for the more complex applications which take more time. So, that by employing this HDFS method, access time could be reduced [5].
(2) **Multiple writes**: As it works on the pattern of write once and read many times, if there is a process which has multiple writes, HDFS method cannot be used [5].

4.1 HDFS Concepts

The following are the concepts of HDFS

(1) **Blocks**: The minimum data which can read or write in HDFS mechanism is known as blocks. By default in HDFS, the block size is 128K. If the files in HDFS are divided into smaller chunks, then it is called block-sized chunks [5].
(2) **Name nodes**: It acts as the master–slave paradigm. Here, name acts as the master and it acts as centralized controller by maintaining all the information in the form of metadata. As metadata is of small size, it can be stored in memory by allowing faster access. Metadata contains information like file permission, name, and location of the block [5].
(3) **Data node**: Data node works under the name node. Periodically, data node reports to the name node, by specifying all the blocks which are maintained by the data node. Data node acts as the commodity hardware and performs same operations like block creation, deletion, and replication [5].

5 Yet Another Resource Manager (YARN)

It is of the core component of the Hadoop, and it is used for the allocation of system resources to the applications which are running in Hadoop. So, main task of Yarn is resource management and job scheduling [6].

5.1 Components of Yarn [6]

1. **Client**: It is used for submitting the MapReduce jobs.
2. **Resources Manager**: It is used to allocate and de-allocate the resource among the clusters.
3. **Node manager**: It is used to monitor entire process on the machine which is on the clusters.
4. **MapReduce Application master**: This component is used to check the tasks running the MapReduce jobs. These tasks are scheduled by resource manager and under the control of node manager [6].

5.2 Benefits of Yarn [6]

(1) **Scalability**: If a MapReduce algorithm runs for 1 time, it can handle 4000 nodes and 40,000 tasks, whereas by using Yarn, we can run 10,000 nodes and 1 lakhs tasks.
(2) **Utilization**: By using Yarn mechanism, utilization of resource will be increased, because here node manager can control a pool of resources rather than fixing no. of resources [6].
(3) **Multi-tenancy**: It is majorly a benefit-related MapReduce. We can run different versions of MapReduce on Yarn [6].

5.3 MapReduce [7]

One of the major features of Hadoop is parallel processing. To utilize this feature, the query should be in MapReduce form.

MapReduce methods have two steps:

  (a) Mapper phase
  (b) Reducer phase.

In mapper phase, a key value pair is provided as input to the mapper and the output of mapper is provided as input to the reducer. The reducer takes this key value format as input and produces the output. The reducer phase will run only after the mapper is over [7].

Steps in MapReduce [7]:

Step 1: As we discussed, MapReduce takes input in the form of pairs and returns a list <key,Value> pair and this pair will be unique.

Step 2: After this <key,Value> pair is applied, we get list of values associated with this pair, then hadoop applies sort ad shuffle techniques on this list and produces the output.

Step 3: Then, the output generated after applying sort and shuffle techniques will be sent to the reducer phase.

Step 4: Then, in reducer phase, a defined function is applied on the list of values and produced the final output of <key,Value> pair.

6 Conclusion

In this paper, a brief discussion about the purpose of bigdata to introduce and how it has been evolutes in the current technology and also discussed about foundation of big data. To handle the big data, an open-source framework named Hadoop had discussed and also how data will be handled using Hadoop and an overview of various modules in Hadoop were discussed.

References

1. https://www.webopedia.com/TERM/B/big_data.html. Last Accessed on 10 July 2019
2. https://www.guru99.com/what-is-big-data.html. Last Accessed on 10 July 2019
3. http://www.dataversity.net/brief-history-big-data/. Last Accessed on 10 July 2019
4. https://www.gutcheckit.com/blog/veracity-big-data-v/. Last Accessed on 10 July 2019
5. https://www.javatpoint.com/hdfs. Last Accessed on 10 July 2019
6. https://www.javatpoint.com/yarn. Last Accessed on 10 July 2019
7. https://www.javatpoint.com/mapreduce. Last Accessed on 10 July 2019

A Review on Datasets and Tools in the Research of Recommender Systems

B. Dinesh Reddy, L. Sarath Chandra Kumar and Naresh Nelatur

1 Introduction

In the past decade, decrease in the cost for storage and computation of data, ease of access to data in web, has made every aspect of our life to be digitised. In the year 2013, Petter Bae Brandtzaeg claimed that 90% of data stored in numerous storage devices are generated in last two years. Exponential growth rate of this digital data has steered to big data era. Practitioners and scientists around the globe are now aspiring to harness these massive volumes of digital data to gain new insights and knowledge. Industries from every field started looking at "data" as the new resource and fuel that can boost economy, and also in combating their competitors. Subsequently, big data steered the development of data-driven products. Emergence of data-driven products has boosted field of recommendation systems (Recsys).

Recommendations systems are now being integrated to every web information system worldwide. Research advances in field of recommendation systems has produced numerous algorithms for generating recommendations. Frameworks for assessing these techniques need to perform evaluation and comparative analysis on datasets. The rational for choosing a dataset is crucial for proper assessment. Accessibility of most appropriate datasets is a bottleneck for practitioners and researchers in this area. This papers objective is to escalate awareness about open/public large-scale datasets for recommendation systems among the research and practitioner's

B. D. Reddy (✉) · L. S. C. Kumar
Department of Computer Science and Engineering, Vignan Institute of Information Technology, Visakhapatnam, AP, India
e-mail: dinesh4net@gmail.com

L. S. C. Kumar
e-mail: sarathlabala888@gmail.com

N. Nelatur
Department of CS and SE, AU College of Engineering (A), Andhra University, Visakhapatnam 530003, India

community. Consequently, it may further put forward state-of-the-art research in this field [1, 2].

2 Research and Datasets

The success of recommendation system is driven by massive volumes of data utilized for generating recommendations. Big data are a buzzword for this decade, but the core concepts are inherited from data mining, a domain of computer science discipline. Further, data mining concepts employ the machine learning algorithms to build a model for prediction task. Research advancements in machine learning have generated thousands of learning algorithms. The fundamental research performed in the area of machine learning has utilized empirical research methods. Analysing the behaviour of these algorithms needs experimental studies, as most of the learning algorithms are too complex to be analysed mathematically. Experimental studies are conducted to gain insight in the behaviour of learning algorithms by training on specific dataset and study of their behaviour. Also, these studies help to measure the performance of an algorithm by doing comparative analysis with the existing state-of-the-art learning algorithms. Thus, open/public datasets are essential for conducting research in field of machine learning and intuitively for recommendation systems.

Competitions and Challenges as Catalysts: In general, research and its evaluation in field of data mining and machine learning witnessed numerous experimental studies using dataset from the repository hosted by university of California, Irvine. However, innovations around the world in twenty-first century are escalated by various crowdsourcing platform. These platforms facilitated to conduct innovative competitions and challenges for many real-world problems. Eventually, such crowdsourcing platform is acting as catalysts in making massive volumes of open/public data available for experimental research studies. Consequently, the field of recommendation systems has also received numerous open/public datasets from various crowdsourcing platform. The "Netflix challenge" has aroused the interest of researchers and practitioners around the globe into the field of recommendation systems. Netflix provided a dataset having 100 million rating given by around 5 million of its customers for 17,700 movies. The challenge is to devise an algorithm that predicts whether a particular customer likes or dislikes a movie, with a prize of $1 million. Also, association for computing machinery (ACM) is hosting conference series entitled ACM recommender systems conference (RecSys) yearly, where participants can access datasets for recommendation system from various domains [3–6]. Other popular competitive platforms to investigate about public datasets are given in Table 1.

Table 1 Delightful crowdsource platform

Competition name	Hosted institute/organization
KDD cup [5]	ACM SIGKDD conference on Knowledge Discovery and Data Mining (KDD)
ECML-PKDD [6]	European Conference on Machine Learning and Principles and Practice of Knowledge Discovery in Databases (ECML-PKDD)
AnalyticsVidhya [7]	Analytics Vidhya, India
CrowdANLYTIX [8]	CrowdANLYTIX, India

3 An Insight on Recommender Datasets

ACM recsys challenge is undoubtedly the front runner in promoting the area of recommendation system into industry and academia [7–9]. It has attracted the interest of researchers, academicians and industrialist worldwide by hosting numerous competitions, workshops, conference since 2007. It being a landscape of real-world data from multiple domains steered the development of recommendation system. The platform has elicited the research challenges in adaptability, quality, scalability and performance of recommendation systems. Datasets published for these competitions are accessible only to participants, and their utilization is limited to this conference works only [10–12]. However, to gain an insight how ACM recsys conference series has harvested innovative approaches that boosted the field of recommendations system, an investigation is done on recsys challenges, and the summary is given in Table 2.

3.1 Open-Source Datasets for Recommender System

Web2.0 aroused new era of hosting and posting content on the web. Large amount of experimental data is made accessible through various repositories. Academic institutes are the front runners in hosting and maintaining public data repositories for scientific study. The research in the field of recommender system seeks real-world data generated from intelligent systems from diverse domains. Furthermore, user preference and behaviour analysis can be efficiently studied from their social media and networks data.

Some of the potential datasets that can be utilized in algorithm analysis and experimental studies in field of recommender systems are given in Table 3 [12, 13]. Performing EDA on the datasets can be useful. Exploratory data analysis (EDA) is the process with many goals such as identifying variables having significant relationships, verifying given data are sufficient or not for making inferences, investigating evidences for given hypothesis and many more. EDA is essential for researchers in formulating, revising research questions, choosing appropriate algorithms and in

Table 2 Summary of ACM recsys challenges

Year	Theme	Task	Dataset description
2018	Music recommendation	Increase engagement by automatic playlist creation	Large number of playlist titles and associated track listings
2016/2017	Social network for job seeker	Predict job posting a seeker will interact	Job seeker profile, job posting and interactions
2015	Social network for business	Predict items a user will purchase in a session	Click events and purchase history of users from numerous sessions
2014	Recsys to optimize user engagement	Predict items that increases user engagement	User ratings tweets on movies
2013	Personalized business Recsys	Predict ratings on business by a user	Users reviews on business with user check-ins
2012	Context-aware Recsys/benchmarking	Predict appropriate user for movies to attain maximum impact on social network	Holistic information about movies

advancing further to attain research objectives. Since all the datasets for recommendation system are of high dimensional, numerous exponential paths can be obtained for analysis [14, 15].

4 Tools of Trade

Usually, the success for e-commerce companies who empowered their information systems with recsys depends upon the tools they employ for generating recommendations. Numerous tools have been developed in the last decade under various open source and proprietary licences in the field of recommendation systems. Table 4 presents most widely used tools in the industry and academia. The selection of tools is based on criteria of popularity and open/free utilization. The notations used in table are CF: Collaborative filtering, KNN: k-nearest neighbour, SVD: Singular vector decomposition and MF: Matrix factorization.

Aforementioned list of tools generates recommendations using a wide range of methods developed from data mining and statistical concepts. These methods by their adopted strategy can be broadly classified into collaborative filtering, content-based filtering and hybrid recommenders. Collaborative filtering (CF) is based on user's behaviour in the past to identify the patterns/association between user and items [Tapestry]. CF uses user's activity such as rating, downloads and user preferences. Advantages of CF are accurate results for majority of cases, support domain-free

Table 3 Recsys datasets in various domains

Domain	Dataset details
Social media and networking	Youtube 8 M dataset-6.1 M videos with 3862 classes 3.0 average labelled (https://research.google.com/youtube8m/index.html) The Flickr 100 k dataset consists of 100,071 images as 40 parts (http://www.robots.ox.ac.uk/~vgg/data/oxbuildings/flickr100k.html) 476 million Twitter tweets datasets consists of 20 million user's 476 twitter posts (https://snap.stanford.edu/data/twitter7.html)
E-commerce review	Amazon24 classes products, 18,133,457 reviews, 79,737,561 ratings (http://jmcauley.ucsd.edu/data/amazon/) Restaurant reviews: 5531 restaurants had total of 52,077 reviews (http://www.cs.cmu.edu/~mehrbod/RR/)
Movies review	Epinions 665 k dataset of 40,163 user's 664,824 ratings on 139,738 movies Ciao dataset consists of 7375 user's 278,483 ratings over 99,746 movies FilmTrust dataset consists of 1508 user's 35,497 ratings over 2071 movies MovieLens 10 M dataset consists of 71,567 user's 10,000,054 ratings over 10,681 movies (https://www.librec.net/datasets.html)
News/research article	Yow dataset consists of 24 user's 10,010 ratings over 647 articles (www.barrolee.com/data/dataexp.htm) 20 newsgroups dataset consists of 20,000 articles collected from 20 different newsgroups (https://www.kaggle.com/crawford/20-newsgroups) CiteULike dataset consists of 5551 user's 204,986 ratings over 16,998 articles (www.citeulike.org/faq/data.adp)
Music	Last.fm dataset consists of 992 user's 19,150,868 ratings over 176,948 items (https://labrosa.ee.columbia.edu/millionsong/lastfm) Million Song dataset consists of 1 M songs and respective ratings (https://labrosa.ee.columbia.edu/millionsong/) Audioscrobbler dataset consists of 20,000 user (https://github.com/topics/audioscrobbler)
Travel/Tourism	Trip Advisor dataset consists of 235,793 reviews (https://nemis.isti.cnr.it/~marcheggiani/datasets/) Expedia dataset consists of 37,670,293 reviews (https://www.kaggle.com/c/expedia-hotel-recommendations) Gowalla dataset consists of 950327 user's 6,442,890 reviews over 196,591 items (www.yongliu.org/datasets/) Foursquare dataset consists of 3112 user's 27,149 reviews over 3298 items (http://www.yongliu.org/datasets/)

Table 4 List of popular tools for generating recommendations

Library	Brief details
Easyrec [15]	*Java that provides personalized recommendations using RESTful Web Services.* Algorithms: Neighbourhood (user CF)
PREA [16]	*Java-based personalized recommendation algorithms toolkit.* Algorithms: Memory-based neighbourhood (user, item CF, Slope-One), matrix factorization (SVD, NMF, PMF, Bayesian PMF)
LibRec [17]	*Java implementation of recommended algorithms.* Rating Prediction: kNN (user, item), matrix factorization (biased, social, trust, Bayesian), SVD (Time, Trust, Reg++); Item Ranking: Personalized ranking (Bayesian, group, social) and LDA
Duine [18]	*Memory-based filtering algorithms written in java.* Algorithm: Pluggable algorithms and dynamic algorithms for collaborative filtering (social filtering) and information filtering
Suggest [19]	*Top-N recommendation engine, implemented as a library.* Rich algorithms for user, item-based CF filtering
Case recommender [20]	*Framework of recommendation algorithms implemented in python.* Rating Prediction: kNN (user, item, user./item attribute), MF (SVD); Item Recommendation: kNN (user, item, user./item attribute), MF using Bayesian ranking, Ensembler of Bayesian personalized ranking
Mahout [21]	Linear algebra-based data mining. Algorithms: Item CF, matrix factorization with ALS, ALS on implicit feedback, weighted matrix, SVD++
MyMediaLite [22]	*C#& Java implementation of recommended algorithms.* Rating Prediction: kNN, modern matrix factorization (biased); Item Recommendation: kNN, modern matrix factorization (weighted regularized, Bayesian)
LensKit [23]	*Java-based toolkit for generating recommendations.* Algorithms: Memory-based user, item CF, Slope-One and SVD with gradient descent for MF
Crab [24]	Python framework for building recommender systems. Algorithms for nearest neighbourhood, user-based similarity
Waffles [25]	Open-source C++ toolkit for recommender. Model-based approaches using machine learning and data mining techniques
Surprise [26]	Python scikit building and analysing recommender systems Matrix factorization (SVD Feature)
GraphLab [27]	Python library. Item based, item content based, factorization, popularity
Recommender Lab [28]	Facilitates infrastructure to generating recommendations in *R* environment. Collaborative filtering, SVD, rule-based association mining
Rival [29]	Open-source Java toolkit for evaluation CF (User/Item), SVD
pyRecLab [30]	Python library for recommendation system. Algorithms: kNN (user/item), SVD, popular

implementation, no need to analyse content, serendipity and ability to capture subtle characteristics. Limitations are cold start, not effective for long tail and difficult to provide reasoning for generated recommendations. Major approaches for CF are neighbourhood and model based. Neighbourhood approach looks to identify the neighbours based on similarity. If the similarity of like mined users is considered, it is known as user-based CF. Conversely, item-based CF considers user preferences. Subsequently, information from neighbours is then used to produce top recommendations. Model-based approach characterizes the data (users, items) to construct model, and the recommendations are generated from the model instead of considering whole dataset. Models can be developed using latent factor, machine learning techniques. Latent factor models generate factors that are inferences from pattern in data, and matrix factorization-based models are proven to be most successful for recommendations systems. *Content filtering* characterizes user/items by developing profiles. Content filtering uses content and descriptions metadata. *LensKit*, *MyMedia* and *Mahout* can be easily integrated with websites for generation of recommendations.

Evaluation Measures: The behaviour of the algorithms and their results are to be properly evaluated using evaluation frameworks. The evaluation of the recommender systems is still an active area of search in spite of many frameworks with numerous metrics are developed [31]. This paper provides preliminary metrics adopted in recommender evaluation studies. Metrics can be classified into error based, accuracy based and ranking based on usage of recommender model. Among these metrics, the confusion matrix, i.e. classification of the outcome of a recommendation of an item to a user, is as follows.

Based on the outcome of the recommendation with respect to Table 5, the performance of the model in generating recommendation through the task of rating prediction is calculated using metrics often categorized as error-based metrics. Much of the early studies in the performance of the recommender models are focused on these error-based metrics. Primary metrics of such category are listed in Table 6.

Metrics for measuring the relative ordering of items in the list are known as ranking metrics. Ranking score is determined by the similarity between orderings of predicted items and those of ground truth discussed in Table 7.

Apart from aforementioned baseline metrics for measuring and estimating the quality of recommendation, additional metrics quantifying various aspects of the recommender systems are also in usage. A few of such metrics are outlined in Table 8.

Higher values for *coverage* metric indicate that recommendations are generated after rigorous exploration of item/product space. Increasing *diversity* without compromising accuracy will ensure that recommendations does not look stupid with repetitions and improve user engagement. High *learning rate* indicates the systems in ability to adapt new data to improve recommendation in a limited time period.

Table 5 Contingency table

	Recommended	Not recommended
Preferred	True-Positive (TP)	False-Negative (FN)
Not preferred	False-Positive (FP)	True-Negative (TN)

Table 6 Error-based measures for evaluation

Metric	Formula	Parameters
Mean absolute error (MAE)	$MAE(R) = \frac{\sum_{(u,i) \in R} |\hat{r}_{ui} - r_{ui}|}{N}$	$\hat{r}_{ui}$ is estimated value, r_{ui} is actual value, N is no. of observations
Root-mean-square error (RMSE)	$RSME(R) = \sqrt{\frac{\sum_{(u,i) \in R}(\hat{r}_{ui} - r_{ui})^2}{N}}$	$\hat{r}_{ui}$ is estimated value, r_{ui} is actual value, N is no. of observations
Normalized MAE	$NMAE(R) = \frac{MAE(R)}{r_{max} - r_{min}}$	$MAE(R)$ is mean absolute error value, r_{max} is the highest error value, r_{min} is the lowest error value
Normalized root-mean-square error (RMSE)	$NRMSE(R) = \frac{RMSE(R)}{r_{max} - r_{min}}$	$RMAE(R)$ is root-mean-squared error value, r_{max} is the highest error value, r_{min} is the lowest error value
Max error	$max - error = \max_{i=1}^{N} \hat{r}_{ui} - r_{ui}$	$\hat{r}_{ui}$ is predicted value and r_{ui} is true value
Precision	$Precision = \frac{TP}{TP+FP}$	TP is no. of true-positive predictions. FP is no. of false-positive predictions
Recall	$Recall = \frac{TP}{TP+FN}$	TP is no. of true-positive predictions. FN is no. of false-negative predictions
F-measure	$F = 2 * \frac{Precision * Recall}{Precision - Recall}$	
Area under curve	$\frac{1}{2}\left(\frac{TP}{TP+FN} + \frac{TN}{TN+FP}\right)$	AUC classifier's ability to avoid false classification
Receiver operating characteristic (ROC)	ROC curve is a graphical plot of TP-rate over FP-rate	True-positive and false-positive rates

Discovery of unknown items to users will make sensible recommendation to users. Lower values of *serendipity* mean over specialization of recommendation algorithm and result in low user satisfaction. If the *throughput* is low, usability of the generated recommendations will be ineffective to the user. Lacking in stability by recommender algorithm will lead to ambiguity in decision-making on recommendations. Reduction in hit ratio and prediction shift indicates that system is *robust* to bitter intentions of some wicked users. Practitioners must ensure good trade-off between the base metrics (accuracy) and aforementioned metrics for an effective generation of quality recommendations.

Table 7 Ranking-based measures for evaluation

Metric	Formula and parameters
Mean Average Precision (MAP)	$AP(x, y) = \frac{1}{x_q^+} \sum_{k=1}^{\left|x_q^+\right|+\left|x_q^-\right|} \text{Prec@k}(y); \left[k \in x_q^+\right] x$ is recalls levels, Prec@$k(y)$ is precision at level k to get output variable y
Precision @k	$KNN(q, y, k) = \mathbb{1}[\text{Prec@k}(q,y) > 0.5], q$—relevant items
Mean reciprocal rank (MRR)	$\left(1, 1/2, 1/3, \ldots 1/\left(1+\left|x_q^-\right|\right)\right)$, x_q^- is top level of recall
Normalized discounted cumulative gain (NDCG)	$\text{NDCG}(q, y, k) = \frac{\sum_{i=1}^{k} D(i) \mathbb{1}\left[i \in x_q^+\right]}{\sum_{i=1}^{k} D(i)} D(i) = \begin{cases} 1 & i=1 \\ 1/\log_2(i) & 2 \le i \le k \\ 0 & i > k \end{cases}$ x_q^+ is the maximum relevant items, $D(i)$ discounted cumulative gain at position i
Half-life utility HLU	$R_u = \sum_j \frac{\max(r_{uj}-\text{default},0)}{2^{(j-1)/(\alpha-1)}} r_{u,i}$: user's actual rating to item ranked at j. Default: default rating value (usually average) α: half-life parameter
Kendall's Tau	$\tau = \frac{C^+ - C^-}{\sqrt{C^u}\sqrt{C^s}} C^u$: no. of pairs of items for which the reference ranking asserts an order. C^+ and C^-: no. of these pairs ranking the correct order and the incorrect order
Spearman	$\rho = \frac{1}{n_u} \frac{\sum_i (r_{i,u}-\bar{r})(\hat{r}_{i,u}-\bar{\hat{r}})}{\sigma(r)\sigma(\hat{r})}$ where $\bar{\cdot}$ and $\sigma(.)$ denote the mean and standard deviation

5 Conclusion and Future Opportunities

Over the past decade, innovations in industries and scientific communities have been intensified with the advent of open innovation process. Open innovation leveraged by crowdsourcing has pushed the frontiers of research. In the near future, open innovation will produce ground-breaking research discoveries for the sustainable growth of human living in all aspects. Currently, its impact is widely spread across numerous industries and academia. In general, deliverables of a scientific research are datasets and articles, illustrating experimental setup and results. Research reproducibility along with scientific licensing is a major bottleneck in acceleration of scientific innovation. Reproducible research is imperative for accelerating progress

Table 8 Additional measures for evaluation

Characteristic	Indication
Coverage	Systems degree of consideration of users/items for recommendation process
Diversity	Measures the dissimilarity between each of the recommended items
Learning rate	Systems rate of learning from new information
Novelty	Discover the unknown items from the user's past preferences
Serendipity	System's ability to generate recommendations are that are positively surprising (novel, unexpectedness) and relevant to users
Scalability/throughput	Systems performance with respect to large number of users/items data
Stability	Internal consistency among list generated by same algorithm on same data
Robustness	Robust to attacks from users to deliberately tamper other recommendations

of science. Comprehensively, reproducibility is defined as sharing of holistic information about experimental setup, dataset, programmatic code and computational results. This sharing content can make others to reproduce the empirical studies and analyse efficiently. A natural phenomenon observed in any successful product's technological advancement is that today's optional feature is tomorrows specifications. Consequently, every intelligent system will be powered by personal assistants to render its services optimally. Personal assistants can be considered as a proactive agent that act as an ultimate recommendation system, which recommends the exactly one right product/action at right time that user seeks in a given context. The field of recommender systems is overwhelming with numerous novel algorithms in the research literature. The level of details provided on these algorithms is apparently inflexible to reproduce. Reproducing, implementing and usage of the work illustrated in the literature are hard as the majority of articles have obfuscated design and analysis details and inconsistent evaluation process. Consequently, understanding and usage of algorithms proposed are becoming impossible. It is essential to reconsider the approach for research documentation, experimentation and evaluation in recommender systems to produce impactful insights and effective progress in extending towards evolution of ultimate recommender systems. It is advisable to adapt the following suggestions in research publications:

1. Using open-access datasets, making datasets publicly accessible in non-proprietary formats, providing references, links about the data sources.
2. Detailed specification with references to computing environment, i.e. employed tools, packages, system software's with versions.
3. Sharing verified, functional and complete code through open licensing on platforms such as Github.com and BitBucket.org.
4. Framework, toolkits for providing correct, complete, consistent code and data associated with it for reproducing efficiently.

Acknowledgements This publication is an outcome of the R&D work undertaken in the project under the Visvesvaraya PhD Scheme of Ministry of Electronics & Information Technology, Government of India, being implemented by Digital India Corporation (formerly Media Lab Asia).

References

1. D.H. Park, H.K. Kim, I.Y. Choi, J.K. Kim, A literature review and classification of recommender systems research. Expert. Syst. Appl. **39**(11), 10059–10072 (2012)
2. F. Ricci, L. Rokach, B. Shapira, Recommender systems: Introduction and challenges, in *Recommender Systems Handbook*, ed. by F. Ricci, L. Rokach, B. Shapira (Springer, Boston, MA, 2015)
3. J. Bennett, S. Lanning, The netflix prize, in *Proceedings of KDD Cup and Workshop*, vol. 2007 (2007)
4. J. Lu, D. Wu, M. Mao, W. Wang, G. Zhang, Recommender system application developments: A survey. Decis. Support. Syst. **74**, 12–32 (2015)
5. https://www.kdd.org/kdd-cup
6. http://www.ecmlpkdd.org/
7. https://www.analyticsvidhya.com/
8. https://www.crowdanalytix.com/
9. https://recsys.acm.org/
10. Martin Robillard, Robert Walker, Thomas Zimmermann, Recommendation systems for software engineering. IEEE Softw. **27**(4), 80–86 (2010)
11. J. Bobadilla, F. Ortega, A. Hernando, A. Gutiérrez, Recommender systems survey. Knowl.-Based Syst. **46**, 109–132 (2013)
12. R. Burke, Hybrid recommender systems: Survey and experiments. User Model. User-Adapt. Interact. **12**(4), 331–370 (2002)
13. J.A. Konstan, J. Riedl, Recommender systems: From algorithms to user experience. User Model. User-Adap. Inter. **22**(1–2), 101–123 (2012)
14. G. Adomavicius, A. Tuzhilin, Toward the next generation of recommender systems: A survey of the state-of-the-art and possible extensions. IEEE Trans. Knowl. Data Eng. **6**, 734–749 (2005)
15. http://easyrec.org/api
16. J. Lee, M. Sun, G. Lebanon, Prea: Personalized recommendation algorithms toolkit. J. Mach. Learn. Res. **13**, 2699–2703 (2012)
17. G. Guo, et al., LibRec: A Java library for recommender systems. UMAP Workshops, vol. 4 (2015)
18. http://duineframework.org
19. http://glaros.dtc.umn.edu/gkhome/suggest/overview
20. A. da Costa, E. Fressato, F. Neto, M. Manzato, R. Campello, Case recommender: A flexible and extensible python framework for recommender systems, in *Proceedings of the 12th ACM Conference on Recommender Systems* (ACM, 2018)
21. S.G. Walunj, K. Sadafale, An online recommendation system for e-commerce based on apache mahout framework, in *Proceedings of the 2013 Annual Conference on Computers and People Research* (ACM, 2013)
22. Z. Gantner, S. Rendle, C. Freudenthaler, L. Schmidt-Thieme, MyMediaLite: A free recommender system library, in *Proceedings of the 5th ACM Conference on Recommender Systems* (ACM, 2011)
23. M.D. Ekstrand, et al., Rethinking the recommender research ecosystem: Reproducibility, openness, and LensKit, in *Proceedings of the fifth ACM conference on Recommender systems* (ACM, 2011)

24. M. Caraciolo, B. Melo, R. Caspirro, Crab: A recommendation engine framework for python. Jarrodmillman Com (2011)
25. M.S. Gashler, Waffles: A machine learning toolkit. J. Mach. Learn. Res. MLOSS **12**, 2383–2387 (2011)
26. H. Nicolas, Surprise, a Python library for recommender systems. http://surpriselib.com (2017)
27. Y. Low, J. Gonzalez, A. Kyrola, D. Bickson, C. Guestrin, J. Hellerstein, GraphLab: A new parallel framework for machine learning (UAI, 2010)
28. M. Hahsler, in *recommenderlab: A Framework for Developing and Testing Recommendation Algorithms* (2015)
29. A. Said, A. Bellogín, Rival: A toolkit to foster reproducibility in recommender system evaluation, in *Proceedings of the 8th ACM Conference on Recommender systems* (ACM, 2014)
30. G. Sepulveda, V. Dominguez, D. Parra, pyRecLab: A software library for quick prototyping of recommender systems. arXiv preprint arXiv:1706.06291 (2017)
31. A. Gunawardana, G. Shani, A survey of accuracy evaluation metrics of recommendation tasks. J. Mach. Learn. Res. **10**(12), 2935–2962 (2009)

Performance Comparison of Different Machine Learning Algorithms for Risk Prediction and Diagnosis of Breast Cancer

Asmita Ray, Ming Chen and Yvette Gelogo

1 Introduction

Breast Cancer is one of the leading causes for demise of woman. According to the recent statistics of World Cancer Research fund (WCRFI) it is the second crucial reason of deaths due to this most exquisite and internecine disease globally [1]. In 2018 reported by the WCRFI two million cases are estimated out of which 626,679 deaths were approximated. Developing countries are mainly predominance by this deadly disease [2, 3].

Breast cancer develops when cell in the breast begin to divide abnormally. Breast can be originates from different parts of the breast. But most common is ductal carcinoma, develops in duct. Duct is a tube which takes part to carry milk to nipple. Another type is lobular carcinoma. Lobular are the glands which are responsible for producing milk. Lobular carcinoma grows in the lobular gland and gradually proliferates to nearby tissue. Spreading of the cancer cell from breast to different parts of the body through blood vessels and lymph vessels is called Metastasis.

The most common traditional method to detect the breast cancer is breast biopsy. The conventional procedure of diagnosing the breast cancer includes mammogram and ultrasound. Machine learning technique is used as alternatives for classification of benign and malignant tumors.

A. Ray (✉)
Dept of Computer Science and Engineering, Vignan Institute of Information Technology, Visakhapatnam, A.P, India
e-mail: ray.asmi@gmail.com

M. Chen
Harbin University of Commerce, Harbin, China
e-mail: mingchen@163.com

Y. Gelogo
Iloilo Science and Technology University, Iloilo, Philippines
e-mail: yvette.gelogo@isatu.edu.ph

Detection of breast cancer mammography plays a crucial role. Magnetic Resonance Imaging (MRI) is most acceptable alternative to mammogram to confirm the existence of tumor. Alergetic reaction to the contrasting agent and skin infection are the main pitfall of MRI test. Early detection of breast cancer can be treated effectively and that can enhance the prediction and survival rate notably [4].

Last few decades several algorithms have been proposed for classification and prediction of breast cancer [5]. Our present work demonstrated the comparison of four classifiers: Support Vector machine (SVM), K-Nearest Neighbors (K-NN), Naive Bayes (NB), Classification and Regression Trees (CART) algorithms. This study endeavored to select the better and most influential data mining algorithms by evaluating the efficiency and effectiveness in terms of accuracy, sensitivity, specificity and precision.

2 Materials and Methods

SVM Classifier:
SVM is a statistical learning theory [6]. It is most popular powerful algorithm due its robustness in machine learning approaches [7]. This classification technique is basically based on the decision planes. Decision planes define the decision boundaries. Decision plane consists of distinct set of objects with different class memberships. This algorithm maps the input space for the non linear data to higher dimensional feature space using non linear mapping and class members are splitted from non-members by constructing hyper plane.

"Margin" concept has been introduced by SVM on either side of a hyper plane. Hyper plane basically isolates the two classes. Maximization of margins is considerably efficacious as it is used for fabricating the largest possible distance between the hyper plane and also the samples on either side. It is used to minimize an upper bound on the expected generalization error.

KNN classifier:
KNN classifier is a significantly widely used classifier, used for classification as well as in regression [8, 9]. This is extensively used for pattern recognition and prediction analysis. This algorithm evaluates the distance between two closest data point from the arrival points. Euclidian distance is broadly used method to assess the distance. The next level is to consider the data with shortest distance. The closest data point is determined by the odd number and it is based on the number of classes i.e. two.

Naive Bayes Classifier:
Naive Bayes Classifier basically uses the concept of Bayesian theorem. It is extremely fast statistical predictive models, frequently utilized in supervised learning [10]. This popular classification technique not only analyzes the relationship of the attributes but also the class belongs to each instance for obtaining the conditional probability. Probability of each class depends on its occurrence in the training data set.

CART Classifier: CART is recursive partitioning method [11]. This algorithm predicts continuous variables or categorical variables. These two variables obtain from the set of continuous predictors or categorical factors [12]. Classification and Regression both tree can be constructed by this algorithm. In this algorithm the data set is split into two sub groups and the process is continued for each sub group until to reach to minimum size of sub group.

3 Experiment

3.1 Breast Cancer Dataset (Mammographic Image Analysis Society)

In this study a benchmark data set Mammographic Image Analysis Society (MIAS) is used for evaluation purpose [13]. This dataset comprises of 700 mammogram images of left and right breast. It is classified into two cases (a) benign (460) and malignant (240).

3.2 Simulation Software

In this study all experiments on the classifiers were conducted using WEKA. Waikato Environment for Knowledge Analysis (WEKA) software is a Java based open source tool. It is used as a ML tool also. In 2006 it was first released to the public under the GNU General Public License [14]. Weka is basically collection of group of different machine learning algorithms which are useful for data preprocessing, classification, regression, association rules and clustering purpose.

4 Performance Results

The original mammograms in MIAS are very large (1024 × 1024 pixels) in size and it deals with large number of noise that degrades the quality of image and that affect in the detection of disease. Preprocessing step removes the unwanted parts that are not required for detection purpose (pectoral muscle and any artifact) [15]. After preprocessing step, analysis of data takes place on the basis of effectiveness and efficiency.

The predictive model has been evaluated by 10 fold cross validation test. In order to train the model and determine the test set this test splits the original set into a training sample.

4.1 Effectiveness

Table 1 represents the effectiveness of all classifiers. Classifiers effectiveness have been measured in terms of correctly classified instances, incorrectly classified instances and accuracy. Figure 1 shows the corresponding results of Table 1.

Efficiency
Table 2 shows the accuracy of the predictive mode that has been measured based on the value of precision, recall, True Positive (TP) rate and FP (True Positive) rate for C4.5, SVM, NB and K-NN.

Confusion matrix is also called as error matrix. It helps to utilize the performance of algorithm. In this matrix instances in a predicted class indicates by row and columns symbolize the instances in an actual class.

Table 1 Performance of classifier

Evaluation criteria	Classifiers			
	SVM	KNN	NB	CART
Correctly classified instances	680	671	665	666
Incorrectly classified instances	20	28	35	34
Accuracy (%)	98.14	96.27	95.99	95.23

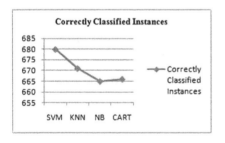
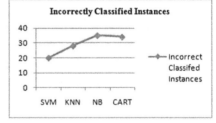
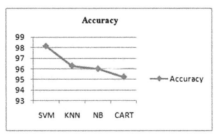

Fig. 1 Performance comparison graph of different classifiers

Table 2 Performance comparison in terms of accuracy measures for SVM, KNN, NB, CART

Classifiers	True positive	False positive	Precision	Recall	F-measure	Class
SVM	0.98	0.02	0.99	0.98	0.98	Benign
	0.95	0.03	0.94	0.95	0.94	Malignant
KNN	0.97	0.06	0.93	0.96	0.95	Benign
	0.90	0.03	0.93	0.90	0.92	Malignant
NB	0.96	0.03	0.99	0.96	0.97	Benign
	0.98	0.04	0.92	0.97	0.94	Malignant
CART	0.94	0.04	0.95	0.94	0.95	Benign
	0.93	0.03	0.90	0.93	0.92	Malignant

5 Discussion

Table 1 represents correctly and incorrectly classified instances and accuracy. SVM classifier obtained highest accuracy (98.14) in other hand accuracy of other classifiers varies between 96.27 and 95.23%. SVM has highest value of correctly classified instances and lower value of incorrectly classified instances compared to other classifiers.

Table 2 shows the efficiency of our algorithm which obtained by analyzing the performance of different algorithm. Highest value has been achieves by SVM for benign class, 98% malignant class has been predicted by NB. FP rate is notably lower for SVM classifier (benign class is 0.02, Malignant class is 0.03). SVM produces the better outcomes from the other classifiers.

Table 3 depicts the confusion matrix that represents the comparison between the actual class and predicted class. SVM classifier able to anticipate 683 instances correctly out of 700 instances (450 benign instances and 233 malignant instances) and incorrect prediction is 17, out of which 10 instances actually belongs to benign class but it correctly detects as malignant class, 7 instances from malignant class but it has been identified as benign class.

The result has been proven that in all respect SVM is the better classifier with lower error rate. It able to exhibits its superiority in terms of efficiency and effectiveness on accuracy and recall for accurately diagnose the breast cancer.

Table 3 Confusion Matrix

SVM	450	10	Benign
	7	233	Malignant
KNN	449	11	Benign
	20	220	Malignant
NB	438	22	Benign
	7	233	Malignant
CART	436	24	Benign
	13	227	Malignant

6 Conclusion

Various data mining and machine learning algorithms have been proposed for analyzing the medical data perfectly. In this paper, to find out the best accurate model for breast cancer diagnosis the performance of various machine learning algorithms have been compared. This study exhibits four algorithms SVM, KNN, NB and CART and their performance have been compared on the MIAS data set. Efficiency and effectiveness of each algorithm has been investigated in terms of accuracy, precision, specificity and sensitivity. SVM has achieved the highest performance based on the simulation result. SVM reaches to the accuracy of 98.14% and it has proven itself as an extensively promising algorithm of breast cancer prediction with regards to precision and low error rate.

References

1. V. Singh, S.P. Lal, Digit recognition using single layer neural network with principal component analysis. IEEE Asia-Pacific world congress on computer science and engineering, Fiji, 1–7 2014
2. Breast cancer statistics. [Online]. Available: http://www.wcrf.org/int/cancer-facts-figures/data-specific-ancers/breastcancer-statistics. Accessed on 25 Aug 2017
3. W. Chen et al., Cancer statistics in China, 2015. CA. Cancer J. Clin. **66**(2), 115–32 (March 2016)
4. J. Heymach et al., Clinical cancer advances 2018: annual report on progress against cancer from the American society of clinical oncology. J. Clin. Oncol. **36**(10), 1020–1044 (January 2018)
5. A. Osarech, B. Shadgar, A computer aided diagnosis system for breast cancer. Int. J. Comput. Sci. Issues. **8**(2) (March 2011)
6. K. Shiny, Implementation of data mining algorithm to analysis breast cancer. Int. J. Innov. Res. Sci. Technol. **1**(9), 207–212 (2015)
7. Y. Rejani, S.T. Selvi, Early detection of breast cancer using SVM classifier technique. Int. J. Comput. Sci. Eng. **1**, 127–130 (2009)
8. G. Williams, Descriptive and predictive analytics. Data Min. with Ratt. R. Art Excav. Data Knowl. Discov. Use R. 193–203 (2011)
9. J.S. Snchez, R.A. Mollineda, J.M. Sotoca, An analysis of how training data complexity affects the nearest neighbor classifiers. Pattern Anal. Appl. **10**(3) (2007)
10. Y.I.A Rejani, S.T Selvi, Early detection of breast cancer using SVM classifier technique. Int. J. Comput. Sci. Eng. **1**(3) (2009)
11. F. Faltin, R. Kenett, Bayesian Networks. Encycl. Stat. Qual. Reliab. **1**(1), 4 (2007)
12. J. Han, M. Kamber, Data mining; concepts and techniques. (Morgan Kaufmann Publishers, 2000)
13. S. Gupta et al., Data mining classification techniques applied for breast cancer diagnosis and prognosis. Indian J. Comput. Sci. Eng. **2**(2), 188–195 (2011)
14. D. Narain Ponraj, M. Evangelin Jenifer, P. Poongodi, J. Samuel Manoharan, A survey of the preprocessing techniques of mammogram for the detection of breast cancer, J. Emerg. Trends Comput. Inf. Sci. **2**(12), 656–664 (2011)
15. G. Valvano, G. Santini, N. Martini et al., Convolutional neural networks for the segmentation of microcalcification in mammography imaging. J. Healthc. Eng. **2019**, 9360941 (9 pages) (2019)

Analysis of DRA with Different Shapes for X-Band Applications

P. Suneetha, K. Srinivasa Naik, Pachiyannan Muthusamy and S. Aruna

1 Introduction

DRA is basically antenna used for the microwave of high frequency having block of ceramic alloy which mounted on a metallic surface. Introducing radiation inside resonator through transmitter which causes boundary to and fro in the walls to bring standing waves. Radiation to space is feasible without metal part. DRAs avoid the losses due to the metallic components by controlling the aspect ratio. DRAs can be made suitable for application in compact wireless devices such that antenna effect is possible by periodic surveying of elements from the capacitive element to inductor like ground plane. Improvement of impedance bandwidth with return loss parameter $S_{11} \leftarrow 10$ dB for use in multiple wireless bands has been the objective of research, and various approaches have been adopted for this includes, the lower quality factor matching of impedance [1–7] and utilization of multiple resonances.

P. Suneetha (✉) · K. Srinivasa Naik
Vignan's Institute of Information Technology (A), Visakhapatnam, AP, India
e-mail: rsss4m@gmail.com

K. Srinivasa Naik
e-mail: nivas97033205@gmail.com

P. Muthusamy
Vignan University, Vadlamudi, Guntur, AP, India
e-mail: pachiphd@gmail.com

S. Aruna
Andhra University College of Engineering (A), Visakhapatnam, AP, India
e-mail: aruna9490564519@gmail.com

© Springer Nature Singapore Pte Ltd. 2020
J. Fiaidhi et al. (eds.), *Smart Technologies in Data Science and Communication*, Lecture Notes in Networks and Systems 105, https://doi.org/10.1007/978-981-15-2407-3_10

2 Antenna Configuration

Geometrical configuration plays a major role in designing DRAs; cylindrical, spherical, and conical shapes are some of the accepted ones for the purpose. Cylindrical geometry is manageable to flexibility of design. In comparison with cylindrical shape, spherical type offers improved gain and better directivity; moreover, the design of spherical shape is dependent only on radius and so it is easier and convenient for fabrication accordingly. The design is implemented using CSTMW studio because it offers ease of implementation and time of execution. A ground plane of size 40 mm × 30 mm × 1 mm is selected as per the size constraints. T-shaped slot having two opposite ends is created by etching on the ground plane. L-shaped and four rectangular slits are created on ground plane for making the operation feasible. A U-shaped micro-strip feed line is also incorporated. This is printed below the Rogers R03010 (lossy) substrate (dielectric constant 11.2) of thickness $t = 0.8$ m. All the three DRAs are designed with Rogers TMM 13i with permittivity $\varepsilon_r = 12.2$, $r = 5.9$ mm, and $h = 4.3$ mm (Figures 1, 2 and 3).

3 Results

To validate the proposed designs, the main performance parameters are return loss, gain, and far field directivity. To obtain results, CST Microwave Studio version 2018 is used for all three proposed antenna structures with time-domain solver. The CDRA antenna exhibits return loss of 39.84 dB; VSWR ratio is 1.020742 at the resonating frequency fr = 9.48 GHz which is shown in Fig. 4a, b; and corresponding gain is 5.57 dB, directivity is 4.98 dB shown in Fig. 4c.

For spherical DRA, the return loss value is 36.02 dB as shown in Fig. 5a; that is, $S_{11} = -36.02$ dB at the resonating frequency fr = 9.11 GHz and the VSWR ratio is 1.03 as shown in Fig. 5b; and corresponding gain is 7.97 dB and directivity is 8.48 dB shown in Fig. 5c.

In Fig. 6a, the conical DRA return loss of 14.02, i.e., $S_{11} = -14.02$ dB, at the resonating frequency fr = 9.87 GHz is observed and the VSWR ratio is 1.44 as shown in Fig. 6b and corresponding gain is 3.54 dB and directivity is 4.92 shown in Fig. 6c and gain plot of spherical DRA is shown in Fig. 6d.

All the dimensions of cylindrical, spherical, and conical DRAs put in Table 1. From Table 2, it may be observed that spherical DRA has return loss −36.02 dB, directivity is 8.48 dB, and gain is 7.97 dB; all these are better than the corresponding values of cylindrical and conical shapes.

Analysis of DRA with Different Shapes for X-Band Applications 79

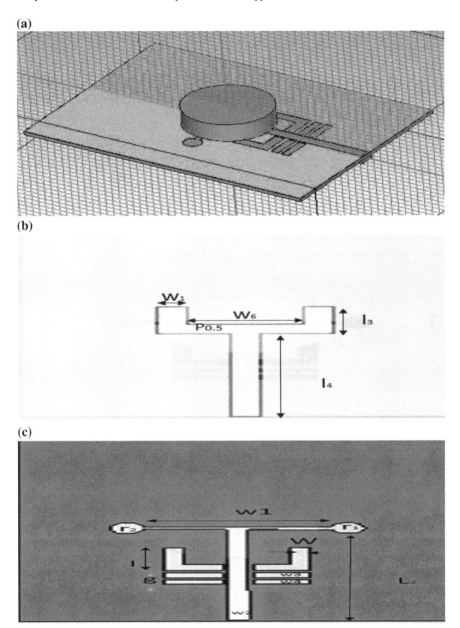

Fig. 1 **a** Perspective view of cylindrical DRA, **b** U-shaped feed line CDRA, **c** top view of CDRA

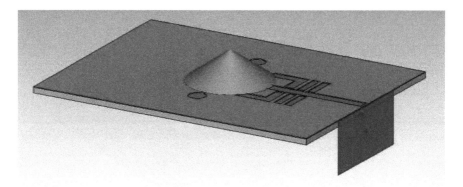

Fig. 2 Perspective view of conical DRA

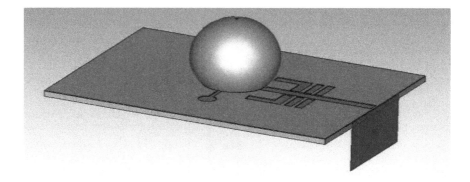

Fig. 3 Perspective view of spherical DRA

4 Conclusion

It has been conclude from the comparison that the spherical DRAs proposed in this paper are the other shaped such as cylindrical and conical in the matter of directivity 8.48 dB, return loss −36.02 dB, and gain 7.97 dB and provides wider bandwidth about 48%. Moreover as it is easy to implement, the spherical-shaped DRAs are opted for less.

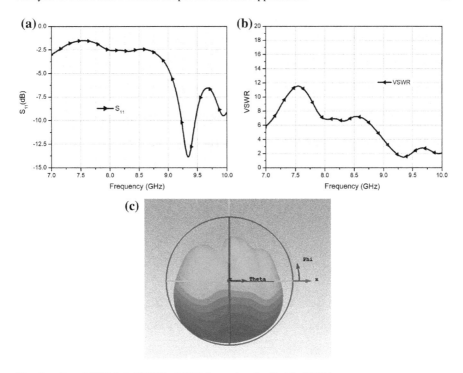

Fig. 4 **a** S_{11} of CDRA, **b** VSWR of CDRA, **c** gain of cylindrical DRA

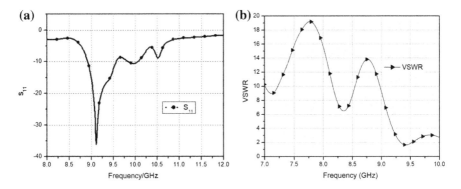

Fig. 5 **a** S_{11} of spherical DRA, **b** VSWR of SDRA

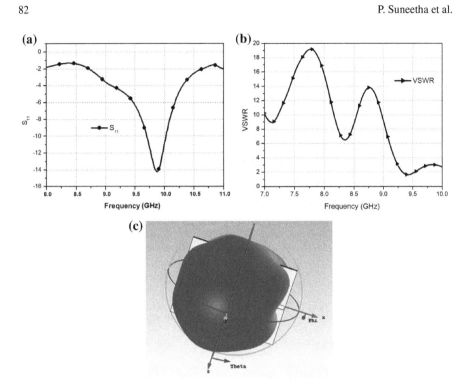

Fig. 6 **a** S_{11} of conical DRA, **b** VSWR of conical DRA, **c** gain of conical DRA

Table 1 Dimensions of CDRA, SDRA, and conical DRA

DRA parameters for all shapes	r	5.9	Composite aperture parameters	l	5
	h	4.3		l1	4
	po s			l2	19.7
Composite aperture parameters	r1	1.2		g	0.6
	w	1.35	Micro-strip feed parameters	w5	2
	w1	12.76		w6	8
	w2	1.5		w7	2
	w3	1		l3	20
	w4	0.6		l4	6

Table 2 Performance comparison of cylindrical, spherical, and conical DRAs

	Cylindrical dielectric resonator antenna	Spherical dielectric resonator antenna	Conical dielectric resonator antenna
Resonating frequency (GHz)	9.48	9.11	9.87
Return loss (dB)	−39.89	−36.02	−14.2
VSWR	1.02	1.03	1.44
Directivity (dB)	5.57	8.48	4.92
Gain (dB)	4.98	7.97	3.54

Acknowledgements This work is supported by DST Science & Engineering research board (SERB) with File: EEQ/2016/000391.

References

1. R. Chair, A.A. Kishk, K.F. Lee, Wide band simple cylindrical dielectric resonator antennas. IEEE Microw. Wirel. Compon. Lett. **15**(4), (April 2005)
2. A. Buerkle, K. Sarabandi, H. Mosallaei, Compact slot and dielectric resonator antenna with dual resonance, broadband characteristics. IEEE Trans. Antennas Propag. **53**(3), 1020–1027 (March 2005)
3. M.S. Al Salameh, Y.M.M. Antar, G. Seguin, Coplanar-waveguide-fed slot-coupled cylindrical dielectric resonator antenna. IEEE Trans. Antennas Propag. **50**(10), 1415–1419 (Oct 2002)
4. R.K. Mongia, A. Ittipiboon, Theoretical and experimental investigations on cylindrical dielectric resonator antennas. IEEE Trans. Antennas Propag. **45**(9), 1348–1356 (1997)
5. K. Srinivasa Naik et.al., Design and analysis of different patch geometry and complementary split ring resonator for X-band applications. Int. J. Recent. Technol. Eng. (IJRTE) **7**(5S4), 858–869 (Feb 2019)
6. K. Srinivasa Naik et.al., Parametric study of vivaldi antenna with different corrugated edges for microwave imaging applications. Int. J. Innov. Technol. Explor. Eng. (IJITEE) **8**(6C2), 283–286 (April 2019)
7. K. Srinivasa Naik et.al., Design, analysis and parametric study of rectangular dielectric resonator antenna arrays. Int. J. Adv. Sci. Technol. **112**, 67–78 (2018)

Android-Based Application for Environmental Protection

Bonela Madhuri, Ch Sudhakar and N. Thirupathi Rao

1 Introduction

Air quality plays a major concern for public health, the environment and ultimately the economy of all the industrialized countries. As health of the citizens living in a country is very important for the development of a country with respect to the industrialization and for better living standards of the citizens of any particular country. The indoor environments are portrayed by a few contaminating sources [1, 2]. Thus, the indoor air quality is an essential factor which is to be controlled to enhance the well-being and solace variables of the tenants as individuals invest of their energy in indoor environments. Today, majority of the people are exposed to artificial environments which need to be controlled by improving the air quality. Several researchers are also trying to find various solutions for the same reasons. The other problems related to these issues are the pollution levels. As the number of vehicles waiting at signals and on roads increases, the pollution that those vehicles generating are also increasing a lot (Fig. 1).

To make thermally agreeable environments with better than average air quality ventilation can be utilized. The iAQ framework which expects to guarantee, freely and precisely observing the indoor air quality in various rooms of a building [3, 4]. It comprises of checking the air quality utilizing remote correspondence. The framework gathers five natural parameters (air temperature, mugginess, carbon monoxide, carbon dioxide and iridescence) from various areas at the same time [5]. Different sensors can be added to screen particular pollutants.

B. Madhuri · C. Sudhakar (✉) · N. Thirupathi Rao
Department of Computer Science and Engineering, Vignan's Institute of Information Technology (Autonomous), Visakhapatnam, AP, India
e-mail: sudhakarcheetirala@gmail.com

N. Thirupathi Rao
e-mail: nakkathiru@gmail.com

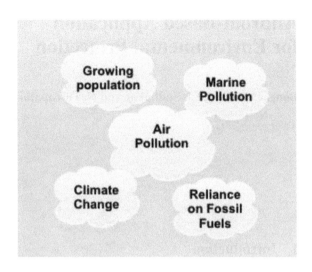

Fig. 1 Top five environmental issues

2 Technical Solution

2.1 Implementation

The iAQ framework is a programmed framework that permits the end client especially assembling administrator to get to different ecological parameters like air temperature, relative stickiness, CO, CO_2 and iridescence.

The iAQ framework is a commonly used unit that can be available in market for cheaper costs with more advantageous and more benefits. It comes generally with a single board and a microcontroller was placed on it. The input and the output units of the device are free and whenever we need to connect, we can connect with them. In most of the cases or scenarios, this unit can serve as both like hardware- and software-based small or mini computer that can be used for several types of applications. Any display unit can be connected to this device and it can be easily monitored by any type of customers or the users at any point of time.

The client can get to the information from the online interface called for iAQ Web created in PHP. Through iAQ Web, the client can see all data about natural parameters. They are appeared as numerical qualities or charts, permits end client to keep up history of indoor air quality for further investigation. With the end goal to permit a speedy, basic, natural and ongoing access a portable application was likewise made with android bolster (Fig. 2).

The noise on the image is filtered and the threshold of each pixel of the image is calculated and the data is used for the further processing. For extracting the background of the image features, the installation and the working of OpenCV and Pi4J libraries in which use counter differentiation algorithm are used. The expert knowledge is in the form of a set of egons where each egon is a monochromic sparse model.

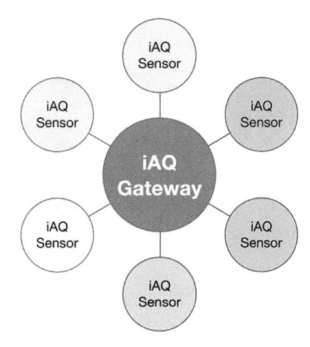

Fig. 2 Wireless sensor network

2.2 Architecture of the Sensor Network

Correspondence is actualized utilizing XBee module, which utilizes the IEEE 802.15.4 system convention, and ZigBee radio protocol (IEEE, 2016). Zigbee is remote correspondence innovation which is of minimal effort and low power dependent on work organizing standard based on 802.15.4 standard (ZigBee Alliance, 2016; Digi International, 2016) [4].

2.3 iAQ Mobile-Mobile Application for Android Devices

It has solid connections to numerical enhancement, giving the field strategies, hypothesis and application areas [5]. AI is now and again connected to information mining, where the last subfield is progressively centered around exploratory information investigation and is known as unsupervised learning. AI is a strategy utilized in the field of information examination to create complex models and calculations that loan themselves to forecast, this is called as prescient investigation in business use (Fig. 3).

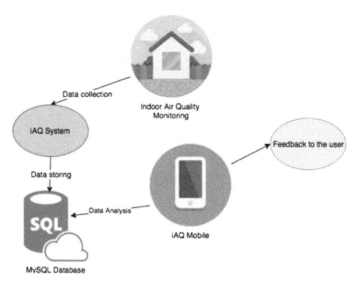

Fig. 3 Application of iAQ android mobile

Fig. 4 iAQ system

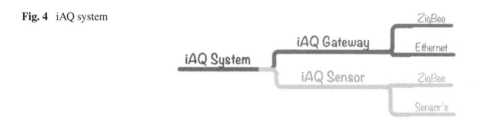

2.4 Hardware and System Architecture

The iAQ framework comprises of at least one iAQ sensor's modules which is utilized to gather and transport air quality information observing in various conditions. The iAQ sensors send the information to the iAQ Gateway which is associated with the Web. Subsequently, it is conceivable to manufacture a secluded framework that can control at least one indoor space all the while (Fig. 4).

The iAQ framework is furnished with various sensors, a handling unit (Arduino MEGA) and a remote correspondence unit that permits work correspondence.

3 Discussion and Results

Figures are the examples that were chosen of relative humidity graphs (Fig. 5), air temperature (Fig. 6) and CO_2 (Fig. 7).

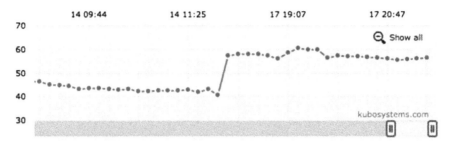

Fig. 5 Relative humidity growth

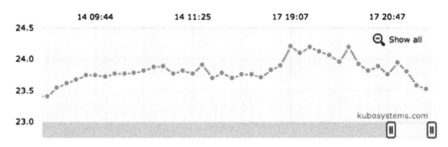

Fig. 6 Temperature chart

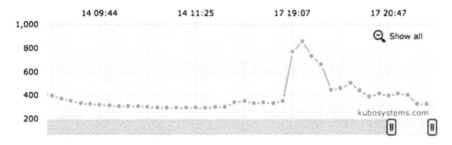

Fig. 7 Graph concentration (ppm) for CO_2

The iAQ system is still in testing phase and the main aim is to make technical improvements, including the proper calibration.

4 Conclusion

The framework intended to show a powerful indoor air quality observing framework to stay away from the dangers of introduction to contaminants in fake condition. The framework was produced by utilizing minimal effort gas sensors and open source innovation, for example, Arduino microcontroller improvement stage. The framework has preferences as far as simplicity of establishment and set up, using remote innovation for correspondence.

References

1. M. Asayama, Guideline for the prevention of heat disorder in Japan. Glob. Environ. Res. **13**(1), 19–25 (2009)
2. D. Chen, K.T. Cho, S. Han, Z. Jin, K.G. Shin, Invisible sensing of vehicle steering with smartphones. Proceedings of the 13th annual international conference on mobile systems, applications, and services, ACM, New York, NY, USA, MobiSys'15, 1–13 (2015)
3. H. Pang, L. Jiang, L. Yang, K. Yue, Research of android smartphone surveillance system. International conference on computer design and applications (iCCDA), vol. 2, 373–376 June 2010
4. https://developer.android.com/guide/topics/sensors/sensors_environment. [Last Accessed on 10-7-2019]
5. K. Fujinami, On-body smartphone localization with an accelerometer. Information **7**, Article No. 21 (2016)

LDA Topic Generalization on Museum Collections

Zeinab Shahbazi and Yung-Cheol Byun

1 Introduction

Text segmentation is one of the important sections in natural language processing. The meaning of segmentation is dividing the document into coherent parts that can be based on words/sentences or paragraphs and finds the similarity between these materials. It prepares the document structure to be useful in text summarization and information extraction [1, 2]. Totally, segmentation is used for removing the repetitive or same meaning sections to get more information from the topics, and it is famous for topic generalization. In this paper, we proposed topic modeling method by using linear regression machine learning algorithms to solve the problem of similarity and find the correspondence between museums in different countries. The dataset which we used through this process is the total information about museums and visitor's comments.

2 Related Works

Try to find information through the document is the first way that comes as an idea to human being which is mostly coming from searching between sentences, but it can make it more complex to detect topics out of that material. To come out of this challenge, proposing machine reader system which is known as SECTOR is defined that it segments the text into relevant parts [3] (SECTOR: a neural model for coherent topic segmentation and classification).

Z. Shahbazi · Y.-C. Byun (✉)
Jeju National University, Jeju City 63243, South Korea
e-mail: yungcheolbyun@gmail.com

Z. Shahbazi
e-mail: z.shahbazi72@gmail.com

Recently, technologies of speech become one of the famous areas in monitoring and it contains high output for user consent. This output is because of analyzing dialogues to catch customer problems and the ways to overcome them (multiple topic identification in telephone conversations).

It is to analyze the structure of document topics or relationship between speeches to show the goal of the speech such as people discussion or customer service dialogues is a significant part to figure out the dialogue, generalize it, and also summarize it (a weakly supervised method for topic segmentation and labeling in goal-oriented dialogues via reinforcement learning) [4].

In late studies in natural language processing, supervised learning has a problem of searching in large amount and labeled data compared with unsupervised learning. To do this, adjusting text segmentation as a document label for every sentence until it ends with segment and contains rather than 727,000 texts from Wikipedia (text segmentation as a supervised learning task) [5].

3 Methodology

In this part, we are trying to present LDA model (latent Dirichlet allocation) to find the museums' activities per year and also find the similarity between different museums. To do this, the proposed method is divided into four main parts: preprocessing, text categorization, topic modeling, and finally similarity measurement between topics. Figure 1 shows the proposed methodology in detail.

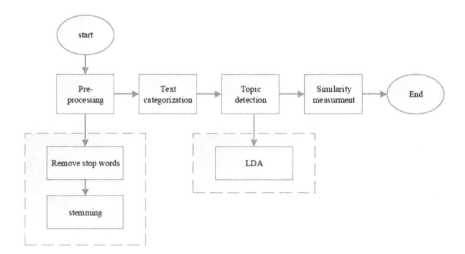

Fig. 1 Proposed system architecture

3.1 Preprocessing

Preprocessing is one of the important steps in the machine learning algorithms. To start the preprocessing section, stop word removal and stemming are used for normalizing the data. Analyzing data can cause problem if the process not done carefully. Preprocessing steps start with extracting terms, transforming, cleaning the data, and end with loading the data.

3.2 Text Categorization

Text categorization is part of natural language processing tasks which is also known as text classification. In this section, unique words extracted from unstructured data types which in this procedure are text documents. Text categorization has two ways that process for extracting information: First step is manually and second is automatically. It usually shows the result based on good quality, but it can be time-consuming. In this paper, we used automatically categorization by applying machine learning ("K-means" and "linear regression") algorithms. The first step toward training the algorithms is feature extraction which converts the text into numerical representation based on vectors. Figure 2 shows the categorization process in detail.

3.3 Topic Modeling

Topic modeling and topic classification are the most popular approaches in the machine learning system and natural language processing. Topic modeling is used to extract the generic information from the document. The proposed system is to recognize the most repeated topics and generalize them into more limited categories based on number of visitors.

3.4 Similarity Measurement

To continue the process of topic detection by using similarity measurement, extracted topics, generalized, and limited them into less topic numbers by categorizing the same meaning topics in one general sub-topic. In case of museum, a comparison of different museums is detected and as a last step recognition between similar topics is processed. Figure 3 represents the similarities between different kinds of museums.

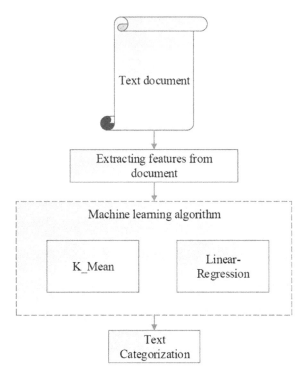

Fig. 2 Text categorization process

4 Experiments

To test the proposed methodology, trip advisor museum reviews used as input dataset which contains 1013 museum information such as museum description, address, entrance fee, and maximum time of visit. This section is to show the detailed results and information about topic generalization between museums.

4.1 Dataset

This dataset is created based on trip advisor information and visitor's reviews. It contains different types of museums, e.g., art, war, history, and car; based on this information, museums activities per year and topic similarity are calculated. Figure 4 shows the comparison between museums that rely on their activities per year and number of visitors.

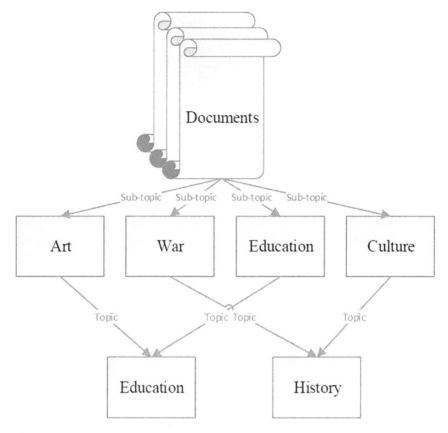

Fig. 3 Preview of topic generalization

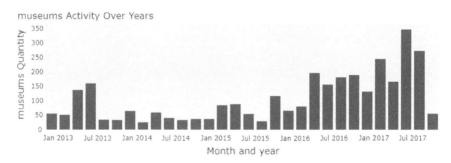

Fig. 4 Yearly preview of museums

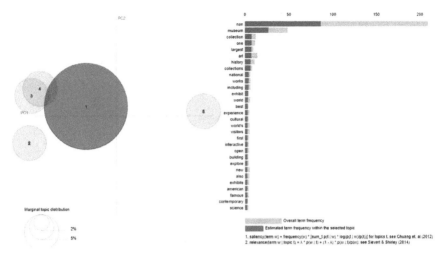

Fig. 5 Representation of topic similarity

4.2 LDA Topic Modeling

Topic modeling is a part of unsupervised natural language processing which display document based on related topics. To automatically organize, understand, search, and summarize large type of texts, topic modeling system has proposed some steps:

1. Finding the hidden part of document.
2. Divide the document based on that information.
3. Apply achieved information to organize, summarize, search, and form predictions.

In this part, document is divided into different topics and after doing the similarity process, topics are generalized into more limited parts, and finally, similarity between these topics is shown in Fig. 5. Table 1 describes the number of topics and percentage of the similarity between topics, running online (single-pass) LDA training, 5 topics, 1 pass over the supplied corpus of 3218 documents, updating model once every 2000 documents, evaluating perplexity every 3218 documents, iterating 50x with a convergence threshold of 0.001000.

4.3 Linear Regression Algorithm

Linear regression machine learning algorithm is used to find the relationship between independent and dependent variables and analyses between topics in document.

Table 2 shows the comparison between latent Dirichlet allocation, similarity feature, and linear regression machine learning algorithm which LDA has 4% higher result than linear regression.

Table 1 Merging changes from 2000 documents into a model of 3218 documents

Topic number	Topic words	LDA similarity PA
1	Museum, history, collection, art, one	0.6
2	Nan, museum, art, one, collection	0.37
3	Museum, history, collection, one	0.2
4	Nan, museum, collection, one, largest	0.4
5	Museum, art, new, collection, one	0.16
Topic number	Topic words	LDA similarity PA

Table 2 Comparison between LDA and linear regression

Algorithms	Topic differences
LDA	0.74
Linear regression	0.70

5 Conclusion

Text segmentation is one of the important aspects of natural language processing which is used for topic modeling. In this paper, we used the combination of LDA similarity feature and linear regression algorithm to find the similarity between museums in different countries. The output represents that proposed system which has 0.4% higher accuracy rate than linear regression one.

Acknowledgements This research was financially supported by the Ministry of SMEs and Startups (MSS), Korea, under the "Regional Specialized Industry Development Program" supervised by the Korea Institute for Advancement of Technology (KIAT).

This work was supported by "Jeju Industry-University Convergence Foundation" funded by the Ministry of Trade, Industry, and Energy (MOTIE, Korea). [Project Name: "Jeju Industry-University convergence Foundation/Project Number: N0002327"].

References

1. F. Author, Article title. Journal **2**(5), 99–110 (2016)
2. F. Author, S. Author, *Title of a Proceedings Paper*, ed. by F. Editor, S. Editor. Conference 2016, vol. 9999 (LNCS, Springer, Heidelberg, 2016), pp. 1–13
3. F. Author, S. Author, T. Author, *Book Title*, 2nd edn. (Publisher, Location, 1999)
4. F. Author, Contribution title. 9th international proceedings on proceedings, Publisher, Location, 1–2 (2010)
5. LNCS Homepage, http://www.springer.com/lncs. Last Accessed 2016/11/21

Roof Edge Detection for Solar Panel Installation

Debapriya Hazra and Yung-Cheol Byun

1 Introduction

Solar panel consists of solar cells which convert light into electricity. Sol or Sun as called by astronomers is the source of light for these solar cells. Solar panels are spread over a large area of the roof where the solar cells conserve renewable energy so that it can be utilized in providing electricity for the home appliances. Installation of solar panels increases access to energy, reduces carbon footprints, and also has low maintenance cost [1]. The amount of energy conserved by the solar panel is directly proportional to the amount of light reaching the panels. Maximum amount of light is emitted by the Sun; therefore, it is important or advisable to align the solar panel with the Sun. So, there is a need to obtain the exact edges of the roof to measure and plan the installation correctly.

To find the boundaries of an object, in our case roof, edge detection is used. Edge detection involves mathematical mechanism which aims to find the points where the image brightness tends to sharpen or change [2]. Connected curves give us the edges of an object in an image which helps in extraction of critical properties or details about the entity. Firstly, noise is removed from an image, then an edge operator is used to detect edges, edges are smoothed using appropriate values of threshold, and then, edge thinning is utilized to remove any factitious points on the edges in an image [2].

Before extracting the edges of the roof tops, it is important to remove any object that is obstructing the boundaries of the roofs. In this paper, we have used Generative Adversarial Network (GAN) for object removal and image completion before

---

D. Hazra · Y.-C. Byun (✉)
Jeju National University, Jeju City 63243, South Korea
e-mail: yungcheolbyun@gmail.com

D. Hazra
e-mail: debapriyah@gmail.com

detecting the edges of the roof. GAN was invented by Ian Goodfellow and his colleagues in 2014. GAN consists of two deep networks that are the generator and the discriminator. The generative model generates new examples or candidates, whereas the discriminative network evaluates them or classifies them as real or fake. In GAN, the generator and the discriminator are trained together. The generator generates real examples and provides it to the discriminator which updates itself in every next step to classify better between real and fake. There is a feedback system which updates the generator whether it could fool the discriminator. This way, two models compete against each other and play a zero-sum game. When the generator is able to fool the discriminator, no changes are required for the generator, but model parameter has to be updated for the discriminator. Similarly, when discriminator successfully identifies real and fake specimen, the generator has to update its model parameter [3]. This is referred to as the zero-sum game.

In this paper, we have used conditional GAN for object removal from the roof images. Deep convolutional generative adversarial network (DCGAN) has been used for image completion after removing the objects, and then to detect the edges of the roof, we have used Canny edge detection method.

2 Literature Review

In the paper, "A generalized Mumford-Shah model for roof-edge detection," the author has used the Mumford–Shah approach for image segmentation to detect edge of roofs with low contrast [4]. They have compared different variations of the Mumford–Shah model and have concluded that the MS model cannot detect roof edges since it does not include second-order derivative terms. They have modified the parameters in MS model to detect edges of the roof images.

"Edge detection using CNN for Roof Images" is the paper where the author has used transfer learning through Visual Geometry Group (VGG-16) Convolutional Neural Network (CNN) to automatically extract the features. The paper uses Robert edge operator on the automatically extracted features to detect the edges of the roof [5].

In the paper "Automated Edge Detection using Convolutional Neural Network," the author points out that using CNN achieves far better results in detecting edges than any traditional or artificial neural network (ANN) methods [6]. The author claims that CNN generates good results also when it is applied to high-resolution images or live images.

"Adversarial Network for edge detection" is the paper where UNET architecture and conditional generative adversarial network has been used to address edge detection problem. This proposed method claims to achieve a speed of 59 and 26 frames per second for an image resolution of [$256 \times 256 \times 3$] and [$512 \times 512 \times 3$], respectively [7]. As the cGAN produces an image that is closed to the real image, the cGAN generator generates edges which are much thinner than the edges obtained from the traditional image processing method.

Edge-enhanced GAN has been used in the paper "Edge-Enhanced GAN for Remote Sensing Image Superresolution" to extract high-frequency edge details in noise-contaminated images [8]. The author claims that the proposed methodology can reconstruct sharp edges and clean image contents that are more realistic and similar to the ground truth [8].

3 Proposed Methodology

In this paper, we have proposed a methodology that uses three steps to obtain the edges of the roof so that the planning for the installation process of the solar panel becomes easier and accurate. The following steps describe our proposed methodology:

1. Removal of the object that is becoming an obstruction in detecting the edges of the roof. To get the clear boundaries of the roof, we use conditional GAN to remove the object from the image.
2. The next step is to restore the background or part of the image from where the object was removed. We used deep convolutional GAN for image restoration.
3. After the object is removed and the image is restored, we detect the boundaries or edges of the roof top. Canny edge detector has been used to detect the edges of the roof.

3.1 Conditional Generative Adversarial Network (cGAN)

cGAN is used to conditionally generate an output. For the generator, the random vector from the latent space is given or conditioned by some extra input in cGAN. The discriminator is also conditioned by giving input image which can be real and fake and an additional input or condition. In short, both the generator and discriminator are conditioned with an additional input in cGAN. We have used cGAN to remove objects like trees, shadow, chimney, and any kind of entity that is guarding the boundaries of the roof. We have developed a user interface, where user can select the pen size and then mark the object to be removed with the pen. Pen size can be changed from 1 to 10. As we can see in Fig. 2, the pen size is 7 and the black part on the left side of the image indicates the object which has been marked by the pen to be removed (Fig. 1).

Fig. 1 Original image

Fig. 2 Marking object with pen that needs to be removed from the image to detect the edges

3.2 Image Restoration Using Deep Convolutional GAN (DCGAN)

DCGAN is one of the network designs for GAN which consists of convolutional layers without max pulling and fully connected layers [9]. DCGAN uses transposed convolution for upsampling and replaces all max pooling with convolutional stride [9]. Except the output layer, batch normalization is used for the generator and the input layer for the discriminator [9]. Figure 3 shows the result of the image completion implemented by DCGAN.

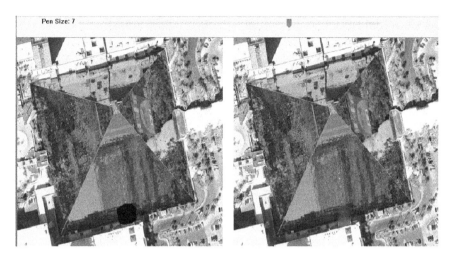

Fig. 3 Image completion using deep convolutional generative adversarial network

3.3 Roof Edge Detection Using Canny Edge Detector

Canny edge detector is a multi-stage detector that applies Gaussian filter first to smooth the image. Then, it finds the intensity gradient and applies non-maximum suppression to remove the fake or extra edges. It applies threshold and detect the edges by hysteresis. After all the preprocessing, removal of object, and image completion, we have used Canny edge detector to detect the accurate edges of the roof for solar panel installation. Figure 4 shows the final result:

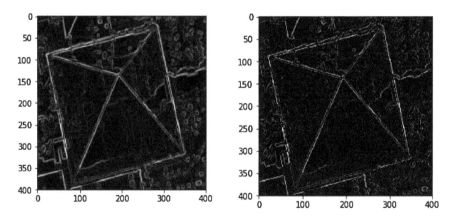

Fig. 4 Roof edge detection using Canny edge detector

4 Conclusion

Our proposed methodology has generated accurate result for roof edge detection which would be helpful to measure and plan for solar panel installation. Usage of cGAN and DCGAN has obtained better results than traditional image processing approaches. We have tested our algorithm with a large image dataset, and the result has been as required and correct 94% of the time.

In future, we would work on automating the whole process of roof edge detection and use machine learning techniques to implement the complete system.

Acknowledgements This research was financially supported by the Ministry of SMEs and Startups (MSS), Korea, under the "Regional Specialized Industry Development Program" supervised by the Korea Institute for Advancement of Technology (KIAT) and "Jeju Industry-University Convergence Foundation" funded by the Ministry of Trade, Industry, and Energy (MOTIE, Korea) [Project Name: Jeju Industry-University convergence Foundation/Project Number: N0002327].

References

1. https://economictimes.indiatimes.com/small-biz/productline/power-generation/benefits-of-rooftop-solar-panels-and-factors-that-further-aid-their-installation. Last Accessed on 10-05-2019
2. https://en.wikipedia.org/wiki/Edge_detection. Last Accessed on 01-05-2019
3. https://machinelearningmastery.com/what-are-generative-adversarial-networks-gans/
4. T.D. Bui, S. Gao, Q.H. Zhang, A generalized Mumford-Shah model for roof-edge detection, in *IEEE International Conference on Image Processing 2005*, Genova, Italy (2005)
5. A. Ahmed, Y.-C. Byun, Edge detection using CNN for roof images, in *The 2019 Asia Pacific Information Technology Conference*, Jeju Island (2019), pp. 75–78
6. M.A. El-Sayed, Y.A. Estaitia, M.A. Khafagy, Automated edge detection using convolutional neural network. Int. J. Adv. Comput. Sci. Appl. **4**(10), 11–17 (2013)
7. Z. Zeng, Y.K. Yu, K.H. Wong, Adversarial network for edge detection, in *2018 Joint 7th International Conference on Informatics, Electronics & Vision (ICIEV) and 2018 2nd International Conference on Imaging, Vision & Pattern Recognition (icIVPR)*, Japan (2019)
8. K. Jiang, Z. Wang, P. Yi, G. Wang, Edge-enhanced GAN for remote sensing image superresolution, in *IEEE Transactions on Geoscience and Remote Sensing* (2019), pp. 1–14
9. https://medium.com/@jonathan_hui/gan-dcgan-deep-convolutional-generative-adversarial-networks-df855c438f. Last Accessed on 15-05-2019

Implementation of Kernel-Based DCT with Controller Unit

K. B. Sowmya, Neha Deshpande and Jose Alex Mathew

1 Introduction

DFT is a powerful tool used in digital signal processing technique. Three main uses of DFT tool are as follows: it used to calculate frequency spectrum of a given signal, used to find frequency response of a system with the help of impulse response, and vice versa. Also it works as an intermediate step in various complex processing methods [1]. Thus, DFT is popular but requires more computing resources and is complex. As compared to DFT, DCT is a better transform which can be used in various signal processing applications as the difference between the two is that DFT utilizes complex exponential functions, while DCT uses only cosine functions (real-valued) [2, 3].

One of the applications where DCT is used extensively is in compression of images. Image compression is a technique that is used to encode the actual image with less number of bits. Minimizing redundancy in an image and efficiently storing its data is the main objective of image compression. Image compression can be classified as lossless and lossy. In lossless compression techniques, the reconstructed image after compression and the original image are similar. While in lossy compression schemes, data loss occurs and introduces compression artifacts. Image which is reconstructed using lossless technique is degraded with respect to original image, and the differences produced by this technique can be considered to be visually lossless [4].

K. B. Sowmya (✉) · N. Deshpande
Department of Electronics and Communication Engineering, R.V. College of Engineering, Bangalore, Karnataka 560059, India
e-mail: kb.sowmya@gmail.com

N. Deshpande
e-mail: nehadeshpande1029@gmail.com

J. A. Mathew
Department of Electronics and Communication Engineering, Srinivas Institute of Technology, Mangalore, Karnataka 574143, India
e-mail: aymanamkuzhy@gmail.com

© Springer Nature Singapore Pte Ltd. 2020
J. Fiaidhi et al. (eds.), *Smart Technologies in Data Science and Communication*, Lecture Notes in Networks and Systems 105, https://doi.org/10.1007/978-981-15-2407-3_14

There have been several developments in lossy compression techniques—JPEG, MPEG and MP3. DCT is used in JPEG encoding where representation of an image is done by 2D array of picture elements called pixels. A grayscale image in JPEG technique is split into smaller blocks of 8 × 8 pixel block to reduce the complexity of computation. In general, in grayscale and color image pixels are represented using bits [5].

2 Discrete Cosine Transform

DCT technique works by separating images into parts corresponding to different frequency. An image/audio input is represented as a block of 8 × 8 blocks of pixels. It provides waveform data in terms of weighted sum of cosine terms. It consists of only even parts of DFT (Fig. 1).

2D DCT equation is defined as

$$H(u,v) = \frac{2}{8}C(f)C(g)\sum_{m=0}^{7}\sum_{n=0}^{7}X(m,n)$$
$$x\cos\left(\frac{(2m+1)i\prod}{16}\right)\cos\left(\frac{(2n+1)j\prod}{16}\right) \quad (1)$$

1D DCT computation is performed by decomposing Eq. (1) into two 8 × 1 blocks

$$Y(k) = \alpha(k)\sum_{n=0}^{(N-1)} u(n)\cos\frac{(2n+1)k\prod}{2N} \quad 0 \leq k \leq N-1 \quad (2)$$

$$\alpha(0) = \sqrt{\frac{1}{N}}; \alpha(k) = \sqrt{\frac{2}{N}}, \quad 1 \leq k \leq N-1 \quad (3)$$

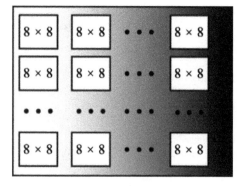

Fig. 1 Representation of an image in terms of 8 × 8 block of pixels

For computing 8 × 8 2D data using 2D DCT technique, first 8 × 1 1D DCT is performed row-wise for all rows which is represented using Eq. (2) followed by 8 × 1 1D DCT performed column-wise for all columns. Transposition memory is used to store results of 1D DCT. Equation (3) represents the constants values used in computation [6].

1D-DCT equation for $N = 8$ can also be written in matrix form as follows:

$$\begin{bmatrix} Y(0) \\ Y(1) \\ Y(2) \\ Y(3) \\ Y(4) \\ Y(5) \\ Y(6) \\ Y(7) \end{bmatrix} = \begin{bmatrix} D & D & D & D & D & D & D & D \\ A & C & E & G & -G & -E & -C & -A \\ B & F & -F & -B & -B & -F & F & B \\ C & -G & -A & -E & E & A & G & -C \\ D & -D & -D & D & D & -D & -D & D \\ E & -A & G & C & -C & -G & A & -E \\ F & -B & B & -F & -F & B & -B & F \\ G & -E & C & -A & A & -C & E & -G \end{bmatrix} \begin{bmatrix} u(0) \\ u(1) \\ u(2) \\ u(3) \\ u(4) \\ u(5) \\ u(6) \\ u(7) \end{bmatrix} \quad (4)$$

Equation (4) can be further simplified by assuming

$$Y(0) = X1 * D + X2 * D + X3 * D + X4 * D \quad (5)$$

$$Y(1) = X5 * A + X6 * C + X7 * E + X8 * G \quad (6)$$

$$Y(2) = X1 * B + X2 * F + X3 * -F + X4 * -B \quad (7)$$

$$Y(3) = X5 * C + X6 * -G + X7 * -A + X8 * -E \quad (8)$$

$$Y(4) = X1 * D + X2 * -D + X3 * -D + X4 * D \quad (9)$$

$$Y(5) = X5 * E + X6 * -A + X7 * G + X8 * C \quad (10)$$

$$Y(6) = X1 * F + X2 * -B + X3 * B + X4 * -F \quad (11)$$

$$Y(7) = X5 * G + X6 * -E + X7 * C + X8 * -A \quad (12)$$

Equation (4) uses A, B, C, D, E, F and G as DCT coefficients—$A = \frac{1}{2}\cos\left(\frac{\Pi}{16}\right)$, $B = \frac{1}{2}\cos\left(\frac{2\Pi}{16}\right)$, $C = \frac{1}{2}\cos\left(\frac{3\Pi}{16}\right)$, $D = \frac{1}{2}\cos\left(\frac{4\Pi}{16}\right)$, $E = \frac{1}{2}\cos\left(\frac{5\Pi}{16}\right)$, $F = \frac{1}{2}\cos\left(\frac{6\Pi}{16}\right)$, $G = \frac{1}{2}\cos\left(\frac{7\Pi}{16}\right)$.

In Eqs. (5)–(12), X1, X2, X3, X4, X5, X6, X7, X8 are defined as

$$X1 = u(0) + u(7) \quad (13)$$

$$X2 = u(1) + u(6) \tag{14}$$

$$X3 = u(2) + u(5) \tag{15}$$

$$X4 = u(3) + u(4) \tag{16}$$

$$X5 = u(0) - u(7) \tag{17}$$

$$X6 = u(1) - u(6) \tag{18}$$

$$X7 = u(2) - u(5) \tag{19}$$

$$X8 = u(3) - u(4) \tag{20}$$

1D DCT is designed by using a pipelined architecture [7]. The architecture mainly consists of three modules—a controller, DCT kernel library and 1D DCT arithmetic unit [8]. The controller block consists of a counter which is used for doing the computations for the given 8 input values for each row. Counter starts from 0 and goes up to 7. The DCT kernel is a library which consists of various DCT coefficients and uses four 8:1 multiplexers for selecting the DCT coefficients for the calculation. Equations (5)–(12) which make use of Eq. (13)–(20) represent 1D DCT arithmetic operation which makes use of adders, subtractors, registers, 2:1 multiplexers and multipliers. Pipelined architecture of 1D DCT is implemented by writing Verilog code in Xilinx Vivado software. Code is written using functions, multiplexers, adders and registers. Inputs values are given to the module where the functionality of addition and subtraction is combined and performed together. Parallel-in parallel-out (PIPO) registers are designed to hold the values obtained after addition and subtraction. Four 2:1 multiplexers are used for selecting either the adder result or the subtractor result. LSB of counter is given a selection line to 2:1 multiplexers. The result of the multiplexer is multiplied with the DCT coefficient obtained using 8:1 multiplexer, selection line to this multiplexer is the counter output, and constant coefficient values are provided as input. Lastly, the output of each 8:1 multiplexer is multiplied with the output of corresponding 2:1 multiplexer and the four multiplier outputs are added to obtain the final results.

3 Modified DCT

Inputs $u(0), u(1), u(2), u(3), u(4), u(5), u(6), u(7)$ each of 8 bits are given to a module where inputs are added and subtracted as $u(0) + u(7), u(1) + u(6), u(2) + u(5), u(3) + u(4)$ and $u(0) - u(7), u(1) - u(6), u(2) - u(5), u(3) - u(4)$. These values are the stored in a register which is triggered by enable control signal, if enable signal is logic HIGH, then the values obtained after addition and subtraction are stored. The register contents are then given as inputs to 2:1 MUX, LSB of the counter present in the controller block acts as selection line to each of the four 2:1 MUX. If LSB of counter is logic LOW, then added inputs are selected from the MUX, and if LSB of counter is logic HIGH, then subtracted inputs are selected from the MUX. DCT coefficients are given as input to the 8:1 MUX, and counter output (3 bits) acts as selection line to each of the MUX. Based on the counter output, one of the DCT coefficients is selected and multiplied with the 2:1 MUX output. The values obtained from each multiplier are added to obtain the final result. The counter value increments on every positive edge of the clock, so at every positive edge of the clock, the new counter value and its LSB are given as selection line inputs to 8:1 MUX and 2:1 MUX, respectively. The counter runs for 8 times as the adder result is to be calculated for the given set of input pixel values. For 24 bit inputs, RGB color with decimal equivalent values are taken as inputs:(6'hFFFFFF) 16777215—white color, (6'hFF0000) 16711680—red color, etc. The final output when divided by 1000 provides the final DCT values which when compared with the original pixel value represent the compressed values (Figs. 2 and 3) [9].

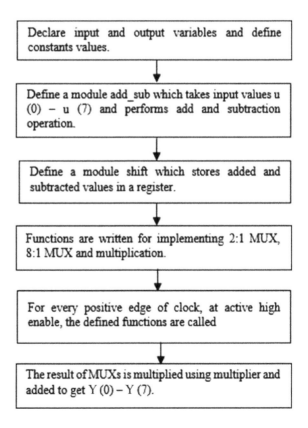

Fig. 2 Modified DCT flowchart

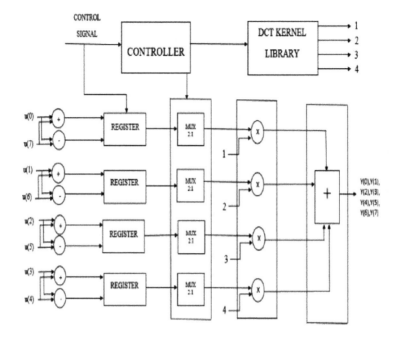

(a) Architecture of DCT arithmetic unit

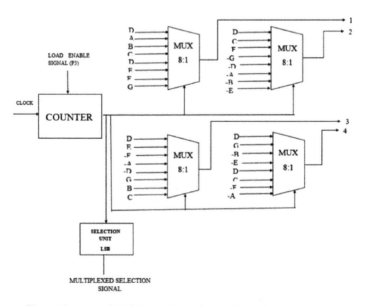

(b) Architecture of DCT kernel library consisting of counter and 8:1 MUX

Fig. 3 Circuit design for generating outputs using **a** DCT arithmetic unit which includes controller, adders, multipliers and DCT kernel library which includes, **b** A 8 bit counter whose 3 bit output controls 8:1 MUX and 2:1 MUX

Table 1 Comparison between the proposed DCT algorithm in [10] and the modified DCT algorithm

Parameters	Proposed 1D DCT algorithm	Modified 1D DCT algorithm
Slice LUTs	696	334
Slice registers	370	304
Min period (ns)	16.29 (Freq. 61.38 MHz)	5.51 (Freq. 181.5 MHz)
Power (W)	2.06 W	154 mW

Table 2 Throughput for FPGA implementation of 8 bit input

Throughput	3 clock cycles

Table 3 Post-synthesis device utilization for 24 bit input

Resource	Utilization	Available	Utilization (%)
Slice LUTs	353	53,200	0.66
Slice registers	307	106,400	0.29
DSP	40	220	18.18
IO	1858	200	929.00
BUFG	1	32	3.12
Power (mW)	159		

4 Results

HDL code is written for implementing 1D DCT equation using the proposed architecture. Code is synthesized using Xilinx Vivado 2014.4 version. Table 1 shows the comparison between the proposed DCT algorithm in [10] and the modified DCT algorithm. Table 2 shows throughput for 8 bit inputs. Table 3 shows post-synthesis device utilization for 24 bit input values. Estimated utilization of flip flops is 1%, LUTs is 2%, DSP48 is 44%, and BUFG is 3% for 24 bit input using the modified DCT algorithm (Fig. 4, 5 and 6).

5 Conclusion

In this paper, a pipeline architecture for computation of 1D-DCT is implemented using Xilinx Vivado software in HDL language which is efficient in terms of power and area [11]. The algorithm includes designing different modules and functions to perform operations such as addition and subtraction of input signals, selecting DCT coefficients based on the counter output, and multiplying multiplexer outputs with DCT coefficients. HDL code is verified and synthesized using Vivado software

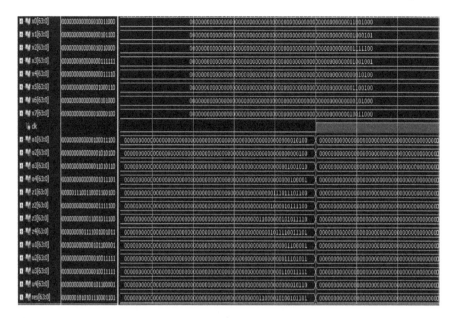

Fig. 4 HDL simulation result for 8 bit grayscale image pixel values using Xilinx Vivado software

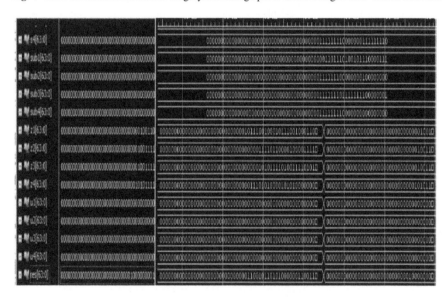

Fig. 5 HDL simulation result for a set of 24 bit RGB image pixel value

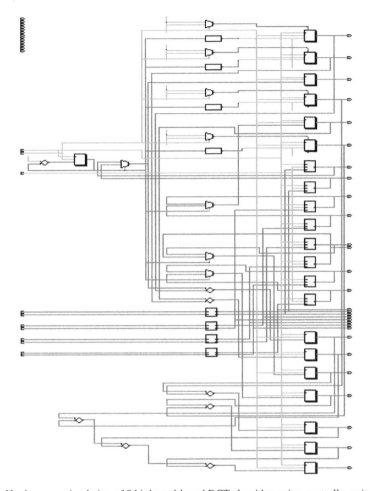

Fig. 6 Hardware co-simulation of 8 bit kernel-based DCT algorithm using controller unit

to provide post-synthesis device utilization and power consumption. HDL code is implemented for 8 bit grayscale image's pixel value and for 24 bit RGB image's pixel value. The modified DCT algorithm requires less area and power as compared to algorithm proposed in [10]. Also the implementation of the modified algorithm for 24 bit input has been successfully done.

References

1. V.K. Sharma, K.K. Mahapatra, U.C. Pati, An efficient distributed arithmetic based VLSI architecture for DCT, in *2011 International Conference on Devices and Communications (ICDeCom)* (IEEE, Mesra, India, 2011), pp. 1–5

2. V.V. Kasturi, Y. Syamala, VLSI architecture for DCT based on distributed arithmetic. Int. J. Eng. Res. Technol. **2**(5), 2278-0181 (2013)
3. K.M. Khatri, S.S. Agrawal, Implementation of discrete Cosine transform using VLSI. Int. J. Eng. Res. Technol. (IJERT) **3**(4) (2014)
4. E.D. Kusuma, T.S. Widodo, FPGA implementation of pipelined 2D-DCT and quantization architecture for JPEG image compression, in *2010 International Symposium on Information Technology* (IEEE, Kuala Lumpur, Malaysia, 2010), pp. 1–6
5. P.D. Chakole, A. Fulsunge, FPGA implementation of pipelined 2D-IDCT for JPEG image decompression. Int. J. Res. Appl. Sci. Eng. Technol. (IJRASET) **5**(VII) (2017)
6. T. Pradeepthi, A.P. Ramesh, Pipelined architecture of 2D-DCT, quantization and ZigZag Process for JPEG image compression using VHDL. Int. J. VLSI Des. Commun. Syst. (VLSICS) **2**(3) (2011)
7. P. Chaturvedi, T. Verma, R. Jain, Design of pipelined architecture for jpeg image compression with 2D-DCT and Huffman encoding. Int. J. Adv. Res. Comput. Sci. Electron. Eng. (IJARCSEE) **2**(1) (2013)
8. A. Tumeo, M. Montero, G. Palermo, D. Sciuto, A pipelined fast 2D-DCT accelerator for FPGA-based SoCs, in *IEEE Computer Society Annual Symposium on VLSI* (IEEE, Porto Alegre, 2007), pp. 331–336
9. H.I. Saleh, FPGA implementation of discrete Cosine transform based image compression encoder, in *3rd Minia International Conference for Advanced Trends in Engineering*, vol. 2 (Minia, Egypt, 2005)
10. H.E. Banna, A.A.E. Fattah, M.W. Fakhr, An efficient implementation of the 1D DCT using FPGA technology, in *Proceedings of the 12th IEEE International Conference on Fuzzy Systems* (IEEE, Cairo, Egypt, 2003), pp. 278–281
11. A. Shams, A. Chidanandan, W. Pan, M.A. Bayoumi, NEDA: a low power high throughput DCT architecture. IEEE Trans. Sig. Process. **54**(3), 955–964 (2006)

An Analysis of Twitter Users' Political Views Using Cross-Account Data Mining

Shivram Ramkumar, Alexander Sosnkowski, David Coffman, Carol Fung and Jason Levy

1 Introduction

The 2016 US presidential election resulted in a polarized country split between two conflicting and unorthodox political views. Traditional polling methods failed to predict the outcome of the election due to the massive political affiliation change among voters. Few studies have been conducted to follow up the political view migration of a large-scale population after the election.

In the USA, more than 70% of the population are social media users [1] and as one of the most popular social network Web sites, Twitter has exceeded a hundred million daily users and nearly one billion total users [1]. This diverse range on one universal platform leads to the representation of almost any political view. Garimella et al. [2] studied the behaviors of US Twitter users and found that a trend of increased political polarization has been formed in the past many years. Therefore, as a more accurate predictor for the upcoming election year and possible future elections, it could be beneficial to look at the change in the opinions of users across a relatively long period of time. In this paper, we conducted a study on the political view migration through following up a selected group of Twitter users over the period of three years.

Our contribution of this paper can be summarized as follows: (1) This is the first work of the same kind to analyze the political views of Twitter users through political score computation. (2) We developed an iterative political score computation algorithm through cross-account data mining on Twitter accounts. (3) We conducted our analysis based on real data on Twitter user behaviors.

S. Ramkumar · A. Sosnkowski · C. Fung (✉)
Virginia Commonwealth University, Richmond, VA, USA
e-mail: cfung@vcu.edu

D. Coffman
Duke University, Durham, NC, USA

J. Levy
University of Hawaii, Honolulu, HI, USA

2 Related Work

Twitter has become a rising social media platform and is used in political campaigns. More research works [3–6] have been done to study political activities on Twitter. For example, Ott [7] used Tweets from Donald Trump as an example to study the impact of social media on conventional journalism. Enli [4] discussed the Twitter strategies of politicians during the US 2016 presidential election campaigns and pointed out that amateurish yet authentic style in social media became a counter-trend in political communication. Garimella et al. [2] studied the behaviors of US Twitter users and found that a trend of increased political polarization has been formed in the past many years. Guess et al. [8] surveyed Twitter users and found worrying amount of individual-level discrepancy that shows the division on political view among population. In our work, we will ascertain the political score of Twitter users based on their Twitter activities and classify them using their score.

3 Analytical Model

3.1 Users' Political Activities in Twitter

Users can take several actions to express their political views in Twitter:

1. *Tweets*: Users can create their own Tweet, visible to all their followers, to express their political views.
2. *Hashtags*: Users can hashtag their Tweets so that it will be visible to other users who browse Tweets with that hashtag.
3. *Retweets*: Users can Retweet someone else's to further disseminate the message. When a user Retweets, it is generally a sign that the user agrees with the view carried in the Tweet.
4. *Comments*: Users can add comments to someone else's Tweets and both the Tweet and the comment. The comments can either agree or disagree with the Tweet.
5. *Likes*: Users can click *like* to show they agree with the content of the tweets.

On Twitter, a key feature for users is to follow any person with a public account which will provide them with updates. It is a common practice that Twitter users follow both people they agree with and ones they disagree with in order to keep up to date react to the Tweets. When a user posts a Tweet, they can choose to address it to a specific user by adding the mentioned symbol @ followed by the target user's username into the Tweet. For example, after a user sends the message "CNN is fake news @real Donald trump" by mentioning.

Donald Trump, the message will be notified to the account *realdonaldtrump*. Another symbol that is frequently used by Twitter users is the hashtag symbol #. Hashtags are often used to label Tweets into certain popular topics areas. For example,

#*NeverTrump* was commonly used by liberal users to label their Tweets against Donald Trump. Hashtags make it easy for users to find all posts that have been thus tagged. Finally, a commonly used feature by many Twitter users is the Retweet function which allows a user to take a preexisting Tweet from a specific user and post it on their own account. In the majority of cases, users what they strongly agree with and would like to share with their own followers. As a result, looking a user's Retweets is a good metric at the rating that user's personal views as the Retweeted message usually comes from a much larger account public account with constant ideals.

To study the political view (liberal or conservative) of Twitter users, we investigated a crowd of randomly selected Twitter users by studying their Tweets connected to political biased users and hashtags. More specifically, we investigated their Retweeting and hashtagging behaviors. The political score of each user and hashtag was computed based on a system of accumulated liberal and conservative credits. Through an iterative cross-count data mining methodology, the political scores were updated until the scores converged. We bootstrapped the algorithm using a set of seed users and hashtags that consisted of prominent politicians and heavily political hashtags.

3.2 Political View Computation Model

We illustrate the Twitter activities relationship in Fig. 1. Circles are users and rectangles are hashtags. The edges between them represent Retweeting or hashtagging. The top round notes with red or blue colors are the labeled seed users. Red color represents conservative view, and blue color represents liberal view. In our case, only prominent political figures such as Donald Trump and Barack Obama are used as seed users. We use a score from 0 to 1 to represent the users' political views. Zero represents complete liberal view, and one represents complete conservative view. Seed users are labeled manually. We use a credit system to estimate the political

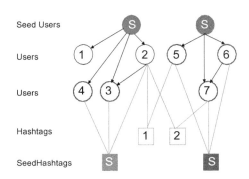

Fig. 1 An example relationship graph of the twitter users and hashtags

score: The credit is accumulated through users' Retweeting and hashtagging behaviors. Let $U = \{u_1, u_2, \ldots, u_n\}$ represent the set of political scores of n users, where $0 \leq u_i \leq 1$ is the score of user i. We use $H = \{h_1, h_2, \ldots, h_m\}$ to denote the political score of hashtag s, where $0 \leq h_i \leq 1$ is the score of hashtag i.

Liberal credits and conservative credits: When a user Retweets another user, commonly it is a sign that he/she agrees with the content of the Retweeted message. Therefore, we use L_i and R_i to represent the liberal credits and conservative credits of user i, respectively. If a user Retweeted from a liberal user, then liberal credit will be accumulated. Similarly, if a user Retweeted from a conservative user, then conservative credit is increased.

When a user tags their Tweet using a political hashtag, it is a sign that the user's tweet conveys the same political view as the hashtag. For example, Tweets tagged with #neverTrump likely convey conservative content. Therefore, we accumulate the liberal credit and conservative credit based on hashtagging behavior as well.

User political score: We compute the political score of a user as follows:

$$u_i = L_i + R_i + 2w_u \qquad (1)$$

where $w_u > 0$ is a parameter to put an initial weight on the computed results to normalize the political scores of users with small credits.

Hashtag political score: We compute the political score of a hashtag based on the number of tweets from liberals and conservatives and as follows:

$$h_j = \frac{\sum_{\forall i \leq n} T_{ij} u_i^{t-1} + w_h}{\sum_{\forall i \leq n} T_{ij} + 2w_h} \qquad (2)$$

where h_j is the political score of hashtag j; T_{ij} is the number of times user I mentioned hashtag j; and u_i is the political score of user i. w_h is an initial weight on the computed results to prevent a large swing of political scores for hashtags with small number of posts from users.

Iterative algorithm: Based on the above method, we derive an iterative algorithm to compute the scores of users and hashtags through iterative computation. The initial scores of users or hashtags at round 0 are set neutral (0.5) except seed users or seed hashtags (they are 0 or 1 if they are conservatives or liberals, respectively). The computations tops when the scores converge.

4 Experiments and Results

4.1 Data Collection

We crawled Twitter information such as users' Retweets and hashtags of 60,491 Twitter users with political interest over the three-year period (2017–2019). These id users included prominent politicians such as Hillary Clinton and Donald Trump. These users are the foundation for the data collection as they are vocal politicians with strong political beliefs. We crawled their followers as our study user base. We also labeled a set of hashtags as seed hashtags under the condition that they are predominantly political slogans with a one-sided meaning. For instance, #never Trump is considered a liberal hashtag and # MAGA is considered a conservative hashtag. We downloaded the most recent Tweet/Retweet activity for each user in our user base in each year. This way we collected our data set for the same collection of users in the year of 2017, 2018, and 2019, respectively.

4.2 Data Analysis

We plotted the political score distribution of a set of users who were able to be scored in all three years (10,343 of them) in Fig. 2. We can see that from 2017 to 2018, there was a notable decline of users with scores of [0, 0.2) (extreme liberal) and (0.8, 1] (extreme conservative) from 2017 to 2018. Many users with scores in the range of [0.2, 0.4) (moderate left) and (0.6, 0.8] (moderate conservative) also shifted closer toward the median; however, in 2019 far more users with moderate liberal and moderate conservative scores exist when compared to either 2017 or 2018. The same year, 2019, also had the largest shift away from neutral values and across all three years. These general observations are also supported in Figs. 3, 4, and 5. Figure 3, a migration graph representing changes in political mindset across users, has a rapid decline in users with extreme political beliefs as seen in the loss of blue or green colors on the graph from 2017 to 2018. Furthermore, Figs. 4 and 5 provide insight into the migration of users across the political spectrum. Notably, from 2017 to 2018, the majority of the extreme liberal users (72.34%) and extreme conservative

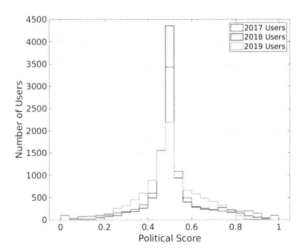

Fig. 2 Political user scores across all three years (2017–2019)

Fig. 3 Political view migration 2017–2018

2018-2019	Switched to:				
	Extreme Liberal	Moderate Liberal	Moderate	Moderate Conservative	Extreme Conservative
Extreme Liberal	1.136%	17.614%	63.636%	17.614%	0.000%
Moderate Liberal	2.628%	20.401%	56.571%	20.401%	0.000%
Moderate	1.621%	17.626%	62.529%	18.223%	0.000%
Moderate Conservative	2.092%	19.710%	60.097%	18.101%	0.000%
Extreme Conservative	2.169%	15.422%	59.759%	18.554%	4.096%

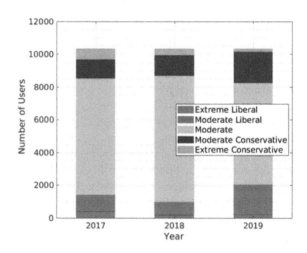

Fig. 4 Political user scores across all three years (2017–2019)

users (75.30%) migrated to more moderate positions; while from 2018 to 2019, a sizeable percentage of neutral (moderate) users became either liberal users (17.626%) or conservative users (18.223%). This can be interpreted as a loss of relevancy of political ideals across Twitter from 2017 to 2018 after the 2016 election and the revival of said relevance in 2019 as a new election season begins.

5 Conclusion

In this study, we designed a novel data mining method to identify US political views by analyzing Twitter data. After analysis, we perceived a pattern that users were gradually shifting from more extreme political positions to moderate ones. The shift toward more moderate stances may be tied to the relevancy of the presidential election (which has the highest turn out of all political elections). As social importance decreased after 2016, users generally devoted less attention to politics and political users; however, as a new election season approaches, users are adopting more polarized stances once again.

References

1. S. Aslam, Percentage of U.S. population with a social media profile from 2008 to 2018
2. K. Garimella, I. Weber, A long-term analysis of polarization on twitter. arXiv preprint arXiv: 1703.02769, 2017
3. M. Ramos-Serrano, J.D. Fernández Gómez, A. Pineda, 'follow the closing of the campaign on streaming': The use of twitter by Spanish political parties during the 2014 European elections. New Media & Soc. **20**(1), 122–140 (2018)
4. G. Enli, Twitter as arena for the authentic outsider: Exploring the social media campaigns of Trump and Clinton in the 2016 U.S. presidential election. Eur. J. Commun. **32**(1), 50–61 (2017)
5. A.D. Segesten, M. Bossetta, A typology of political participation online: How citizens used twitter to mobilize during the 2015 british general elections. Inf. Commun. Soc. **20**(11), 1625–1643 (2017)
6. S. Waisbordand, A. Amado, Populist communication by digital means: Presidential twitter in Latin America. Inf. Commun. Soc. **20**(9), 1330–1346 (2017)
7. B.L.Ott The age of twitter: Donald Trump and the politics of debasement. Crital. Stud. Media Commun. **34**(1), 59–68 (2017)
8. Guess, K. Munger, J. Nagler, J. Tucker, How accurate are survey responses on social media and politics? Polit. Commun. **36**(2), 241–258 (2019)

The Amalgamation of Machine Learning and LSTM Techniques for Pharmacovigilance

S. Sagar Imambi, Venkata Naresh Mandhala and Md. Azma Naaz

1 Introduction

Biomedical literature and clinical reports provide a natural ground to detect and analyze the adverse effects of drugs on a big scale. However, making use of such an enormous quantity of information requires the design of efficient and automatic tools which can help human experts in identifying the drug and its adverse events.

World Health Organization (WHO) defined this process as Pharmacovigilance which detects, assesses, monitors and prevents the adverse effects of drugs. It is very crucial to measure risks of a drug as the usage of drugs is rapidly growing.

An abnormal symptom or disease associated with a medicinal product, i.e., drug, is termed an adverse event [1]. An associated drug may not be considered related to adverse events and is, therefore, a normal drug, and then, they are considered to have a causal association.

Manual identification is a very difficult process, and the human experts appreciate the automatic tools to perform this analysis. Availability of enormous biomedical datasets and medical literature contains tags like named entities and ICD codes that provide information to apply automation in Pharmacovigilance [2].

This research aims at applying deep learning and text analytics tools for identifying the relation between the drugs and its effects. Our method follows two approaches: First approach is finding the casual association between drug and drug reactions as normal or adverse. The second approach is applying LSTM to label each drug in the document as a culprit drug or not.

S. Sagar Imambi (✉) · V. N. Mandhala
Koneru Lakshmaiah Education Foundation, Green Fields, Vaddeswaram, Andhra Pradesh, India
e-mail: simambi@gmail.com

Md. A. Naaz
Andhra Medical College, Visakhapatnam, Andhra Pradesh, India

© Springer Nature Singapore Pte Ltd. 2020
J. Fiaidhi et al. (eds.), *Smart Technologies in Data Science and Communication*, Lecture Notes in Networks and Systems 105, https://doi.org/10.1007/978-981-15-2407-3_17

2 Survey of Literature

Association rule mining is utilized in identifying drug–adverse event associations from the medical documents or case reports. Using this analysis, a knowledge base was constructed which can be further used for future identification [3]. Using aggregate numbers of reports for finding the new drug–adverse event is studied by several researchers. They observed that rule-based solutions are more appropriate for detecting the adverse reactions of a drug/drugs [4, 5] Proportionality analysis of reporting data contains relevant data for detecting drug adverse effects [6]. This idea is opened for further research. One of the extensions was Bayesian confidence propagation neural network (BCPNN) algorithm. It has influenced the disproportionality measure of features [7].

Improvement in drug abnormality detection makes use of stronger predictors (multiple strength of evidence). Using logistic regression, the trends in time and geographic spreads are observed [8]. Collective analysis of reports used dependency of associations present in one report to another and considers less attention to the strength of individual reports.

In the case of the deep vein thrombosis pulmonary embolism [9], it is found that the statistical NLP was better to use in extracting drug–AE pairs. To extract the relationship between drug and adverse event from scientific case reports, Java Simple Relation Extraction (JSRE) was used [10].

Medical Subject Headings (MeSH) helps in finding drug and adverse events association. MeSH provides a list of drugs and its information. We can detect the drug–event association with well-defined adverse event term. Leveraging the acceptable association limits the approach [11, 12]

The researchers observed recurrent convolution neural network as classifier improved the accuracy rate in classifying the medical documents [13, 14]. JB_LSTM was a variation in LSTM for the detection of adverse drug reactions (ADRs) in electronic health records (EHRs) written in Spanish. Sara Santiso et al. developed a joint AB-LSTM network, to consider reinforce lexical variability. They proposed the use of embedding which is created using lemmas. When compared with supervised feature extraction approaches, this approach showed better performance [15].

Conditional random field (CRF) is one of the techniques found for discovering associations for Pharmacovigilance from social media [16, 17] Several researchers have also studied the recognition of an association between chemical compounds and drugs. Grego et al. proposed a chemical entity recognition approach using ChEBI ontology. They used conditional random fields for discovering lexical similarity and chemical terms [18].

Health-related Web sites provide a vast amount of data in terms of consumer product feedbacks and reviews. This datum is rich with information like drugs, its side effects, reactions, adverse reaction. Machine learning classifiers discover the patterns hidden in this information [9].

3 Methodology

This paper shows the implementation of two different methods to find the association between a drug and adverse drug events. The first method is to discover a casual association between drug and drug adverse reactions. The output of this method is the rate of association between a drug and drug reaction and whether that relationship is normal or adverse. Figure 1 shows the architecture of this process. The second method is predicting culprit drugs. In this method, a culprit drug is differentiated from normal drugs using LSTM classifier.

The MEDLINE dataset was groped into 70% and 30% for training and testing, respectively. Ensemble machine learning classifiers are used for the first phase, and LSTM is used in the second method. Validation scheme is similar in both the methods and explained in the next section.

3.1 Ensemble Machine Learning Model for Identifying the Association Between Drug and Drug Reaction

The main steps in this process are

- Data collection.
- Feature extraction.
- Generating level 1 classifier learning model.
- Generating level 2 classifier learning model.

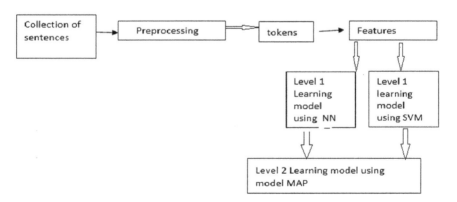

Fig. 1 Architecture to find normal or adverse relation between drug and event

3.1.1 Data Collection

The corpus consists of two types of sentences collected from documents. One is the sentences with drug and adverse effects, and second type is the sentences with drug and normal reactions. The corpus includes the annotated drug, the medical conditions, for example, stomach pain as a medical condition and 'sporlac' as a drug. This corpus should be manually annotated by human experts. The MEDLINE dataset of 12,250 records was grouped into 70% for training and 30% for testing.

3.1.2 Preprocessing

Initially, the sentences collected from the documents had preprocessed by removing noisy data and stop words. Then, each sentence is divided into one gram tokens.

3.1.3 Feature Set Extraction

To identify the association between a drug and its effect on the extraction of relevant terms from the sentence is highly recommended. Drug and medical condition is annotated so the phrases used in the sentence before drug and after drug are identified as features. The drug and the medical conditions are also added to the feature set.

The following features describe each report or document.

- Drug features:

  Include words of related drugs tagged as drug

- Medical condition features:

  Include the words related to event or effect.

- Postdrug or Predrug:

  The three tokens after drug keyword are annotated as postdrug and before the drug keyword is predrug.

- PostMc or PreMc:

  The three tokens after medical condition keyword are annotated as PostMc and before the medical condition keyword is PreMc.
  In the same manner, all the features are collected from the tokens.
  Ex:
  'After taking restyle, the patient felt Numbness in one side of body.'
  The words from the above sentences like 'patient,' 'felt' are PreMc and 'in,' 'one,' 'side,' 'of,' 'body' labeled as PostMc and 'after,' 'taking' are labelled as Predrug and 'patient,' 'felt' are labelled as Postdrug.

  For the dataset, three tokens located next to the drug and medical condition are taken as features. The threshold for number tokens around varies with the minimum length of the sentence. The number of tokens located next to drug and medical condition is considered as training data for further analysis. The number of tokens should balance the trade-off between overfitting and bias. The feature set was built with these generated 1-g tokens.

An ensemble machine learning model was applied to reduce the generalization error. The architecture has two important levels of the learning process. In first level, ANN was applied and then followed by the SVM classification learning model. MAP hypothesis was applied after training each classifier with fivefold cross-validation method. The evaluation of method is presented in Sect. 4.

3.2 Culprit Drug Prediction Approach

This approach generates a model to classify a drug as a culprit drug or non-culprit drug by using a machine learning-based classification model LSTM. The selected training data for learning model include all the relationships in terms of features.

The approach follows the following steps:

- Data collection.
- Feature extraction.
- Generating learning model.

3.2.1 Data Collection

A dataset of around 8400 cases were generated from PubMed [9]. The below query was given for Abstracts search from PubMed.

'Adverse effects Case Reports drug English [lang]'

There were nearly 10,600 drugs that were included in this dataset. All of the drugs were labeled to be a culprit drug or non-culprit drug by the human experts. From the experimental result, it was observed that 41% of the drugs were classified as suspect drugs. Then, the 70% of dataset was allocated to training and 30% was allocated as test data.

The query result of PubMed is shown in Fig. 2.

3.2.2 Feature Set Extraction

From the 8400 case reports, features related to each drug were extracted and the dataset is formed. RNN is used to extract features (tokens) from sentence through first LSTM layer. Next, two LSTM layers are trained for tagging parts of speech tags for each token other than the drug token and named entity recognition using the drug corpora. These two layers are trained alternatively so that the knowledge learned from the previous layer can be enhanced by the next layer.

Fig. 2 PubMed result for the query

3.2.3 Learning Model

Finally, the 20 features are used at the dense layer of LSTM model to classify the given drugs from the sentence as culprit drug or non-culprit drug.

The architecture of the LSTM model is shown in Fig. 2.

Each LSTM layer in the model captures previous token information of a sequence in its memory as shown in Fig. 3. It computes each token of the input sequence (x_1, $x_2, \ldots, x_n$) and transforms it into a vector form (y_t).

The LSTM is designed such that it can remember the sequences of tokens from the previous state. The internal structure of the LSTM cell is shown in Fig. 4.

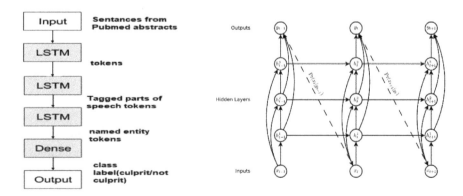

Fig. 3 LSTM architecture

Fig. 4 LSTM memory cell architecture

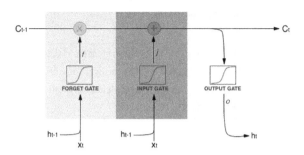

LSTM is a special kind of RNN where hidden states are replaced by memory cells to capture long-term dependent information from previous states. This feature of LSTM is suitable for capturing the drug and its adverse events.

The internal function of the cell starts receiving input information through the input gate. Input gate decides which tokens are going to be stored. Forget gate tells about the tokens which are not going to be stored. So this gate decides which tokens are removed from the sentences.

In our case, it reduces unnecessary information which in turn reduces the dimension of the feature set. The output gate activates the LSTM block and provides the final output, which activates the final output of the LSTM layers at time stamp 't.'

The equations for the states of three cells and the final output of hidden node are

$$C'_t = \tanh(wc[h_{t-1}, x_t] + b)$$
$$C_t = ft * c_{t-1} + i_t * C'_t$$
$$H_t = o_t * \tanh(c_t)$$

where i_t is an input gate, and o_t is the output gate. We pass these two C_t and H_t to the next time step and repeat the same process. The features were generated from the dataset, after analyzing sentences for identifying culprit drugs. A total of 15 features were retrieved to represent all the necessary information from the dataset. Some of the features are listed in Table 1.

4 Experimental Results

This section presents the ML and DL structure used in both methods discovering a casual association between drug and drug adverse reactions and predicting culprit drugs. The output of the first method is the rate of association between a drug and drug reaction and whether that relationship is normal or adverse. The second method shows how culprit drug is differentiated from normal drugs using LSTM classifier.

The evaluation of both the methods was performed on 30% of the respective dataset (test data). The testing of discovering a casual association between drug and

Table 1 Retrieved features from LSTM layers

S. No.	Feature	Remarks
1	Drug name	The drug name matched with the drug index of MeSH
2	Medical condition	The medical condition matched with the drug index of MeSH
3	Drug offside effect	The Mc tokens of after drug
4	Drug count	
5	Single-drug flag	
6	Multi-drug flag	
7	Single MC flag	
8	Multiple MC flag	

Table 2 Confusion matrix of ensemble machine learning model for finding normal/abnormal relation between drug and medical condition

Predicted		Actual	
		Normal	abnormal
Normal		3440	210
Abnormal		630	970

Table 3 Comparison of ML models

Algorithm	Precision	Recall	Accuracy
ANN	88	82	81
SVM	90	83	83
Ensemble	94	85	84

drug adverse reactions was performed on 5250 records out 12,250 documents. The test data size is 5250. Features were created as mentioned in the above sections. Table 2 shows the confusion matrix presenting the accuracy and precision–recall of the ensemble classifier.

The performance of ANN, SVM and an ensemble model is given in Table 3. The ML ensemble model was able to classify 78% of the association between a drug and its effects in the test dataset as normal. The model exhibits 84% of accuracy in the prediction of drug and its effect association as normal or abnormal (Fig. 5).

4.1 Culprit Drug Prediction Approach

For testing culprit drug prediction, 2500 records were used from the extracted dataset. This datum included the documents which are not part of training data. The confusion matrix for the LSTM model is shown in Table 4. Our model discovered culprit drugs from the test dataset with 76% accuracy. The precision and recall of the model are 73% and 90%, respectively.

Fig. 5 Comparison of machine learning models for finding normal/abnormal association between a drug and medical condition

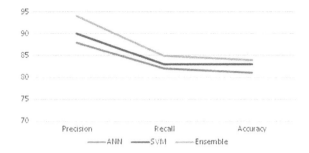

Table 4 Confusion matrix of LSTM model

Predicted		Actual	
		Culprit	Non-culprit
	Culprit	1250	464
	Non-culprit	136	650

5 Conclusion

Medical industry is facing a lot of challenges in processing the increasing volumes of data being generated by the various sources like journals, articles, social media and digital libraries. This research aims at applying deep learning and text analytics tools for identifying the relation between the drugs and its effects. Our method is based upon two approaches: First approach is finding the casual association between drug and drug reactions as normal or adverse. The second approach is applying LSTM to classify each drug present in the document as a culprit drug or non-culprit drug. The first approach yielded a precision of 0.86 and recall of 0.83 in establishing causality between drugs and drug effects on the testing dataset. Subsequently evaluating on the testing dataset of 2500 records, the culprit drug classification successfully predicted culprit drugs and showed significant precision as 73% and recall as 75%. These approaches showed the promising results in finding the culprit drugs and its associated abnormal medical conditions.

References

1. https://en.wikipedia.org/wiki/Adverse_event. Last Accessed on 10-4-2019
2. A. Maitra, K.M. Annervaz, T.G. Jain et al., A novel text analysis platform for pharmacovigilance of clinical drugs. Procedia Comput. Sci. **36**, 322–32 (2014)
3. Séverac, et al. Non-redundant association rules between diseases and medications: an automated method for knowledge base construction. BMC Med. Inf. Decis. Mak. **15**(29), 1–6 (2015)
4. A. Bate, M. Lindquist, I.R. Edwards, S. Olsson, R. Orre, A. Lansner et al., A Bayesian neural network method for adverse drug reaction signal generation. Eur. J. Clin. Pharmacol. **54**(4), 315–321 (1998)

5. H. Gurulingappa, A.M. Rajput, A. Roberts, J. Fluck, M.H. Apitius, L. Toldo, Development of a benchmark corpus to support the identification of adverse drug effects from case reports. J. Biomed. Inf. (2012)
6. S.J.W. Evans, P.C. Waller, S. Davis, Use of proportional reporting ratios (PRRs) for signal generation from spontaneous adverse drug reaction reports. Pharmacoepidemiol. Drug Saf. **10**(6), 483–486 (2001)
7. S. Yeleswarapu, A. Rao, T. Joseph et al., A pipeline to extract drug-adverse event pairs from multiple data sources. BMC Med. Inf. Decis. Mak. **14**(13) (2014)
8. O. Caster et al., Improved statistical signal detection in pharmacovigilance by combining multiple strength-of-evidence aspects in vigirank. Drug Saf. **37**, 617–628 (2014)
9. M. Torii, S.S. Tilak, S. Doan, D.S. Zisook, J.W. Fan, Mining health-related issues in consumer product reviews by using scalable text analytics. Biomed. Inform Insights (2016)
10. H. Gurulingappa, A.M. Rajput, L. Toldo, Extraction of potential adverse drug events from medical case reports. J. Biomed. Semant., pp. 3–15 (2012)
11. R. Winnenburg, N.H. Shah, Generalized enrichment analysis improves the detection of adverse drug events from the biomedical literature. BMC Bioinf. **17**, 250–255 (2016)
12. H. Ayaydin, H. Bozkurt, Spasmodic torticollis associated with sertraline in a child and an adolescent. Turk. J. Pediatr. **57**, 109–111 (2015)
13. S. Lai, L. Xu, K. Liu, J. Zhao, Recurrent convolutional neural networks for text classification, in *Proceedings of the Twenty-Ninth AAAI Conference on Artificial Intelligence* (2015)
14. T. Munkhdalai, F. Liu, H. Yu, Clinical relation extraction toward drug safety surveillance using electronic health record narratives: classical learning versus deep learning. JMIR Public Health Surveill. **4**(2) (2018)
15. S. Santiso, A. Perez, A. Casillas, Exploring joint AB-LSTM with embedded lemmas for adverse drug reaction discovery. IEEE J. Biomed. Health Inf. (2018)
16. A. Nikfarjam, A. Sarker, K. O'Connor, R. Ginn, G. Gonzalez, Pharmacovigilance from social media: mining adverse drug reaction mentions using sequence labeling with word embedding cluster features. J. Am. Med. Inf. Assoc. (2015)
17. R. Zazo et al., Language identification in short utterances using long short-term memory (LSTM) recurrent neural networks. PloS one **11**(1) (2016)
18. T. Grego, F. Pinto, F.M. Couto, LASIGE: using conditional random fields and ChEBI ontology. Association for Computational Linguistics (2013), pp. 660–666

An Artificial Intelligent Approach to User-Friendly Multi-flexible Bed Cum Wheelchair Using Internet of Things

Bosubabu Sambana, Vurity Sridhar Patnaik and N. Thirupathi Rao

1 Introduction

A wheelchair is one of the fundamental requirements in any clinic or hospital. It effectively transports an individual starting with one area then onto the next. These patients, for the most part, have limited developments because of their infections or the shortcoming caused because of their ailments. Such patients need to utilize a wheelchair to move to start with one spot then onto the next. Patients may require moving because of reasons, for example, the need for natural air, expecting to visit restrooms as well as to clean themselves. Wheelchair fills this need as it is modest and most proficient gadget accessible. The patient, who is unable to move because of his/her infections, needs to move from the wheelchair to bed or vice versa. For this reason, they need an external support. The present venture proposes structuring wheelchairs as shown in figure, which could be changed over into beds or vice versa utilizing mechanical linkages [1].

An AI machines analyzes categorical information into various block identities, and predict and improvement is required, immediately intelligent agents directly communicate internal working operations. John McCarthy authored the term in 1956 as part of software engineering worried about causing PCs to carry on like people. It

---

B. Sambana (✉)
Department of Computer Science and Engineering, Viswanadha Institute of Technology and Management, Visakhapatnam, Andhra Pradesh, India
e-mail: bosukalam@gmail.com

V. S. Patnaik
Department of Mechanical Engineering, Visakha Technical Campus, Visakhapatnam, Andhra Pradesh, India
e-mail: srisaipatnaik@gmail.com

N. Thirupathi Rao
Department of Computer Science and Engineering, Vignan's Institute of Information Technology, Visakhapatnam, AP, India
e-mail: nakkathiru@gmail.com

© Springer Nature Singapore Pte Ltd. 2020
J. Fiaidhi et al. (eds.), *Smart Technologies in Data Science and Communication*, Lecture Notes in Networks and Systems 105, https://doi.org/10.1007/978-981-15-2407-3_18

is the investigation of the calculation that makes it conceivable to see the reason and proceed. An artificial intelligence is not quite the same as human being thinking itself and research since its accentuation on calculation and is not the same as computer science and engineering on account of its emphasis on sensitivity, thinking, and activity. It makes machines more intelligent and progressively helpful. It works with the assistance of artificial neurons (artificial neural system) and logical scientific approach.

AI has advanced to the point in offering genuinely reasonable advantages in huge numbers of their applications. Most important artificial intelligence zones are expert systems, natural language processing, speech recognition, robotics, and sensors technology, computer vision and scene recognition, intelligent computer-aided technology, and neural computing. In these, expert system is a quickly developing innovation that is hugely affecting different fields of technological life. The different procedures applied in intelligent reasoning are neural network, fuzzy logic, evolutionary computing, and hybrid artificial intelligence [2].

Intelligent machine owning has the points of interest over the common knowledge as it is increasingly lasting, steady, and more affordable, has the simplicity of photocopying and extending, can be archived, and can play out specific errands a lot quicker and superior to the person.

IoT is a system of internet-empowered items, and cloud services communicate with these multiple objects. It is where items around us will have the option to associate with one another in the existing environment. The IoT will make existence where any of the items is associated with the Internet and speak with one another with predictive object analysis to streamline utilization of assets to build the nature of resources offered to individuals and limit the operational expenses of qualitative services.

We propose that IoT will be taken as a measure to help in the increase in yield of the healthcare environment. Here, the development of intelligence-based systems for the patient user-friendly support sector has to concentrate. The system monitors and provides mobile/SMS alerts based on IOT with real-time monitoring environmental parameters under supervision, if any changes indentify of patient current health conditions like body temperature, Heartbeat, pulse rate, blood pressure etc. Users can view all operations on mobile application through directly and virtually mode.

It includes mobile inspection device provides virtual contact between both parties, and one proposed Physical device connected to IoT equipment, and continusiloy data receiving from various devices and all devices are connected into one platform, statistics gaining unit, Cloud Storage Servers, and the structure be able to repeatedly bring together existing Bed cum chair parameter from the environment [3].

It automatically gives intelligent response by using parameters and presents a continuous health monitoring with line graphical representation analysis in support of the Clients or Users to identify with the necessity of the various parameter, this resolve to permit to update information to control and doctors monitoring the different devices using the mobile application.

2 Related Work

This strategic undertaking is to structure a solid emergency clinical or hospital bed for use in India that can be produced locally. With the culmination of this task, we trust that our bed configuration will be generally utilized all through medical centers in India. We trust that by locally producing the bed, the cost of the bed will be definitely decreased and take into account reasonable present-day human services to be given to a critical bit of India's populace. At first, we accumulated writing from various assets with respect to existing bed structures. With endless supply of existing structures, we created plan frameworks to figure out what made a solid emergency clinic bed and which characteristics our preferred bed would utilize and afterward created according to CATIA drawings utilizing CATIA V5.

When fruitful finishing of outlines was created, we made a working model of our bed and after that ran cross-investigations for auxiliary strength, stress obstruction, and other mechanical properties to guarantee the bed's comfort. The group's definitive objective of structuring a solid local emergency clinic bed for India will be skillful through the resulting conclusion of the associated objectives [4].

- To examine all current bed items and figure out which is of the highest caliber in connection to cost.
- To manufacture our very own plan for use in India dependent on the highest caliber beds.
- To fabricate a model of our bed continually monitoring.
- To decide and encourage the general nature of the emergency hospital beds.

Durability: It incorporates the life expectancy of the bed under ordinary working conditions, extending from 2 years to 20 years. Toughness considers the material(s) the bed edge is planned from, the unwavering quality of the sort of engines utilized in electronic beds, and the straightforwardness and cost of reparability among different capacities.

Safety: The security of each bed was controlled by the material determinations, consistence according to CIS and FDA standards, stress analysis on the weakest areas of the comfortable beds, and the capacities given if there should be an occurrence of a patient emergency.

No difficulty of Manufacturing: Simplicity in manufacturing is a noteworthy thought in bed plan for various reasons, most strikingly the resultant net revenue that would result from the creation proportion, number of laborers required, improvement in one of a kind parts, and so on. Manufacturability was controlled by the reproducible segments on each bed, the trouble for getting together, the evaluated number of laborers to finish each bed, and time for manufacturing each bed [5].

Cost: Bed expenses were considered in two measurements. The main measurement considered was the real cost of getting the bed together and buying though the subsequent component was the expense of shipment to the emergency clinic. So the proposed bed must be financially savvy which means the get-together cost and buy cost and shipment cost must be important minimum for purchasers.

Simple Operation: Today's quick-paced world requires facilitating forms, however, much as could reasonably be expected, and in the packed Indian air, there is brief period to squander while working in the emergency clinic. One of the approaches to decrease the medical attendant to patient time proportion is to make the beds simple to work and simple to figure out how to work [6]. Albeit electronic beds have a favorable position in light of the fact that the engines can move quicker than any human, the hand-turned beds ought to be easy to utilize and as speedy and effective as well.

Electric Functions: Although the vast majority of the American clinic beds we examined had electric capacities to work them, some Indian medical clinics use hand-wrenched beds. The consideration of electric capacities in beds today makes it simpler for patients to accomplish more solace as they themselves can frequently alter their own beds to the patients' very own support levels.

Easy Transportation: Proposed BED cum Chair will move one place to another place easily because user portable friendly device, these device very useful in surgeries, Medical emergency, and Room change along with any other requirements, this model extremely suitable for hospitals, Bed rest persons and paralysis patients. Transportation therefore considered not only the ease to move a particular bed, but also the space required to move each bed.

Design of BED

Figure 1 shows the design model of the "Form Irresolute Couch." The upper portion of the bed is the fabrication of three individual parts of size 2ft x 2ft by joining the hinges between them which enables the rotational movement of the parts. The rack and pinion mechanism is placed under the middle part, and pinion is clamped and arrested for the horizontal and vertical movement of the pinion and only enables the rotational movement. The pinion meshes with the rack which is attached to the links at its ends. The links were attached to the upper portions of the bed. A handle is fixed into the pinion which supports the operation of the pinion [7] (Fig. 2).

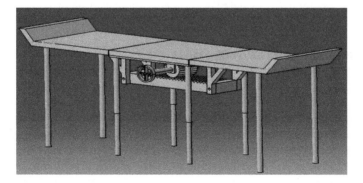

Fig. 1 Form irresolute couch BED outer view

Fig. 2 Proposed designed BED cum chair side view

2.1 Components of the Proposed BED

Lying portion of the bed

Figure 3 shows the "lying portion or upper portion" of the bed. It is made of mild steel material frames of individual dimensions 2ft × 2ft. The three individual parts are assembled with the hinges in between them which enable the rotational movements of the portions.

Pinion

The model shown in Figs. 3, 4 is the "pinion," which is the crucial part of the operating mechanism. The pinion is made of cast iron/aluminum material. The size of the pinion is 100 mm outer diameter and hole of 20 mm inner diameter, which is fixed under the middle portion of the lying portion of the bed.

Rack

The model in Fig. 5 is "rack." It is the crucial part in the operating mechanism. It is

Fig. 3 Lying portion of the bed the model

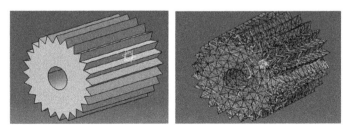

Fig. 4 Operating mechanism of pinion

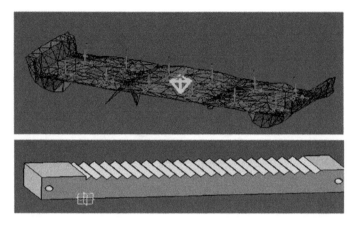

Fig. 5 Operating mechanism of rack

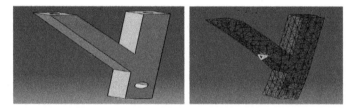

Fig. 6 Supporting link

made up of cast iron/aluminum material. The length of the material is 850 mm, and the width of the material is 70 mm. There were two holes at the two ends of the rack to fix the supporting links. It is supported at its base, and it is always in mesh with the pinion.

Supporting links

The model in Fig. 6 is "supporting links"; it is also an important part in the operating/working mechanism. It is made up of mild steel material; it is in the shape of "Y". It is fixed at the both the ends of the rack and is joins to the first and last portion of the lying portion of the bed. It consists of two legs one with length of 100 mm and another leg of length 60 mm and 30 degrees with the first leg (Table 1).

2.2 Proposed Static BED

- At present, hospitals are using the bed as shown in figure
- Here, in this bed, whole bed is tend to be in constant/static position
- This bed was manufactured as a whole and single unit which is fixed part

Table 1 Required materials

Sl. no	Name	Type of material	Quantity no's
1	Frame	Mild steel square pipes	8
2	Body	Mild steel round pipes	8
3	Wheels	Tubeless tires	2
4	Rack	Cast iron	1
5	Pinion	Cast iron	1
6	Sheets	Plastic	3
7	Ball bearings	Ms	2
8	Mobile application	Software development	3
9	IOT equipments	Hardware development	3

- Using of this bed may cause discomfort/inconvenience to the people like senior citizen, lame/infirm, blind.

3 Proposed/Dynamic Bed Cum Wheel Chair

IoT device can be continuously connected to multiple mobile devices through mobile applications. Artificial intelligent systems scans patient pulse rate and temperature with heart beat rate, etc., and they give responsive data to send representative person, such that person verifies all graphical and imaginarily operation make handy for lame/infirm people. A simplified mechanism with an effective cost is introduced i.e., structure changing bed to replace traditional hospital beds. The proposed system as shown in Fig. 7 is "Form Irresolute Couch" [8].

In this proposed bed, it consists of mainly four parts.

- Upper or head portion assembly
- Lower or leg portion assembly
- Middle or static portion.

3.1 Operating (rack and pinion) mechanism

Upper portion assembly is a part which is designed as per dimensions 2ft. × 2ft., which is fixed to the middle portion with the help of rotating clamps, and it moves upward while working. Lower portion assembly is a part which is designed as per

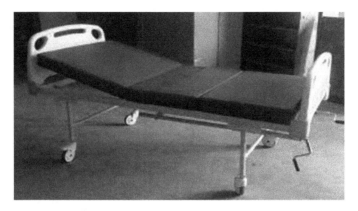

Fig. 7 Existing hospital bed

dimensions 2ft. × 2ft., which is fixed to the middle portion with the help of rotating clamps, and moves downward while working. Middle portion is also called as static portion of proposed bed design, because it is fixed to the body of the bed.

Operating mechanism is unique mechanism which is rack and pinion mechanism, here pinion is fixed and allowed only to move in rotational direction, rack is fixed to the supporting links which are attached to the leg and head portions, and here rack is always in meshed position with the pinion [1].

3.2 Working of Proposed Bed Cum Chair

At first, the bed is in mean position that is in lying position as shown in Fig. 8.

Fig. 8 Mean position OFD bed

In lying position, the three portions of lying portion of the bed will in flat position by rotating the handle which is fixed in the pinion, the pinion will rotate and make the rack move to slide to and fro. If the pinion is rotated in anticlockwise direction with the help of handle fixed in it, then the rack which is in meshed position with the pinion will move toward right direction [7].

The working procedure of bed lever rotation of pinion in mate with rack causes the movements of segments of bed.

- At first, the bed is in mean position that is in lying position.
- In lying position, the three portions of lying portion of the bed will in flat position.
- By rotating the handle which is fixed in the pinion, the pinion will rotate and make the rack move to slide to and fro.
- If the pinion is rotated in anticlockwise direction with the help of handle fixed in it, then the rack which is in meshed position with the pinion will move toward right direction.
- By moving toward right, the supporting links will drag leg portion to the downside and lift the head portion and bed becomes wheel.
- If the pinion is rotated in clockwise direction, then the rack which is in meshed position with the pinion will move toward left direction.
- By moving toward left, the supporting links will drag head portion and lift leg portion to the mean position of the bed.
- Thus bed tends near exchange to wheel chair and vice versa (Fig. 9).

Applications and Future scope

- When patient or user can stay on this chair, then all existing body-scanning sensors, temperature, heartbeat, pulse, and many more sensors can automatically detect and continuously scan the patient body.
- If any emergency occurs, then it automatically sends message alerts to existing users through PDA devices and mobile phones to existing users, all operations monitored through mobile application.
- Form Irresolute Couch can be used in general hospitals, i.e., government hospitals suitable in India.

Fig. 9 Bed converted to wheelchair

- It can be used where patients of any disease who are unable to move from one place to another place. It helps the patients to move from existing bed by converting it into wheel chair.
- It can be useful for the lame/disabled persons, to move in local places.
- It can be used as both normal structure as well as smart attachment structure suitable for condition of the place, area, and purpose.
- It uses Internet of Things as much as possible.
- Mobile usage operation can be achieved.
- Devices like CPR and pulse monitor can be applicable.
- Attachments like saline cum blood bottles and packets holding stands can be applicable.
- Remote control operations can be enabled for blind people, and AI analyzers will control all those functionalities and communicate frequently if necessary action required, but calculate complete functionalities continually go on.

Result:

- Subsequent to achievement of proposed objectives with the production, industrialized, and maintain of our proposed bed cum wheelchair design, we ran a number of tests to guarantee that our bed was certainly artificial in a way which may possibly gather manufacture and FDA procedures.
- The majority significant design gives full relax with stress free environment.

4 Conclusion

An attempt is made to design and fabricate a Form Irresolute Couch. The design of the Form Irresolute Couch ensures that dynamic principles explained on rack and pinion mechanism are applied practically with support of AI and Internet of Things operations by using mobile applications.

References

1. P. Axelson, A guide to wheelchair selection paralyzed veterans of America, 1st edn. (Library of Congress Cataloguing in Public Data, Washington, 1994)
2. J. Kauzlarich, Wheelchair caster shimmy II: damping. J. Rehabil. Res. Dev. **37**(3), 305–315 (2000)
3. R.A. Cooper, T.A. Corfman, S.G. Fitzgerald, M.L. Boninger, Performance assessment of a pushrim activated power assisted wheelchair. IEEE Trans. Control Sys. Tech. **10**(2), 1063–1072 (2002)
4. R. Cooper, Wheelchair selection and configuration, 1st edn. (New York, Springer Publishing Company, 1998)
5. D. Jolly, Wheelchair transfer, Proceedings, 15th IEEE Mediterranean electrochemical PES winter meeting, Columbus, 170–178, 2010

6. A. Winter, Mechanical principle of wheelchair design. Int. J. Mech. Eng. Technol. **7**(2), 261–265 (2010)
7. D. Chakrabarti, Indian anthropometric dimensions for ergonomics design practice, (Ahmedabad, National Institute of Design, 1997)
8. P. Axelson, A guide to wheelchair selection paralyzed veterans of America, 1st edn. (Washington, Library of Congress Cataloguing in Public Data, 1994)

BOSUBABU SAMBANA has been working as Assistant Professor in the Department of Computer Science and Engineering and IPR & MHRD Institute Innovation Cell Coordinator at Viswanadha Institute of Technology and Management, Visakhapatnam, affiliated to Jawaharlal Nehru Technological University Kakinada, Andhra Pradesh, India. He completed Master of Computer Applications and Master of Technology (CSE) from JNTU Kakinada. Currently, he is pursuing AMIE from Institute of Engineers of India (IEI), Kolkata, India.

He filed 21 Indian Patents and among them 8 were published in Indian Patent Journal, Govt. of India. He also wrote one text book named as "Fundamentals of Information Security" and published 3 book chapters in Springer Nature. He secured 3 Copyrights which were granted by Copyright Office, Government of India and 1 Trademark granted. He published 38 research papers in various reputed (UGC approved, Scopus-10/ SCI-2/Web of Science-3) international and national Journals, magazines and conferences. He filed 4 designs and 1 Trademark.

He participated in 14 international and 3 national conferences. He guided 5 UG academic projects. He participated 14 international and National workshops and 7 FDPs. He secured 9 Certifications from NPTEL - IISc/IITs on different specifications like Algorithms for Big Data, C++ Programming, Introduction to Research, Wireless Sensor Ad-hoc Networks, Internet Architecture, Internet Security and Computer Architecture, Artificial Intelligence, Cloud Computing and Block Chain etc. He has 8 years of experience in teaching and research and also has good knowledge on Artificial Intelligence, Astronomy, Cosmology, Space Research, Future Internet Architecture & Technologies, Block chain, Cloud Computing and Internet of Things along with academic subjects etc. Currently, he is the active member of IEEE, IAENG, AMIE, ISA, CSI, ISC, IMS, CSTA, IACSIT, SDIWC, EAI, the IRED, UACEE, Indian Science Congress, LiveDNA, ORCID.

Mr.Bosubabu has intensive editorial experience (Editor-in-chief for International Journal of Research and Development in Engineering Sciences (IJRDES-2582-4201)). He is honored with "Outstanding Researcher Award 2019" by International Institution of Organized Research, "Young Scientist Award 2020" by South Asian Educational Research Awards Society and "Outstanding Scientist Award 2020" by VDGOOD Professional Association.

Dr.Viruty Sridhar Patnaik, working with as Professor and Department of Mechanical Engineering, Visakha Technical Campus, Visakhapatnam, Jawaharlal Nehru Technological University Kakinada, Andhra Pradesh. He has professionally 18+ years Teaching and Industrial and Research Experience. He has published 12 research papers in various National and International Journals. He guided 20 UG and 14 M.Tech academic Projects. He participated 10 international and National workshops and 5 FDPs. He is the active member of IEEE, IAENG, ISC, IMS, CSTA, IACSIT, SDIWC, EAI, the IRED, LiveDNA and ORCID.

A Study on Pre-processing Techniques for Automated Skin Cancer Detection

Netala Kavitha and Mamatha Vayelapelli

1 Introduction

One of the most common types of cancer in the world is skin cancer. When this cancer has been diagnosed early, the patient will survive. Early diagnosis of skin cancer is therefore very important to reduce the mortality rate. According to the World Health Organization, approximately 132,000 melanoma cases occur globally each year. Malignancy (basal, squamous cells, Markel cell carcinomas) is considered to be malignant melanoma [1]. Two major types of cancer are skin cancer. It is highly curable if melanoma is diagnosed in the early stage. Among white-skinned populations, melanoma is the most common cancer and is the lesion among skin layers that contain essential pigments. The treatment of pigmenting skin lesions is made through a computer-aided skin cancer test. Due to the complexity and complications of human understanding, automated assessments of dermoscopic photos must be chosen as a field in research. The high performance of the device has allowed doctors to treat pigmented skin damages in order to avoid misdiagnosis. Pre-processing, segmentation and classification are common for early detection of skin cancer.

The pre-processing can be split into enhancing the image, restoration of images and removal of hair in the detection of skin cancer. Each phase comprises various methods mentioned in this article. The pre-processing techniques are selected depending on the methods for the automation system. Imaging scaling, the transformation of colour space and the enhancement of contrast are some of the most common image enhancement technique as well as Gaussian, speckle noise filters [2] and medium and median filters are some of the filters used in image restoration and blur restore and noise restore are image restoration techniques and morphology

---

N. Kavitha · M. Vayelapelli (✉)
Vignan's Institute of Information Technology, Visakhapatnam, Andhra Pradesh, India
e-mail: India.netala.kavitha@gmail.com

N. Kavitha
e-mail: mamatha.vaylapelli@gmail.com

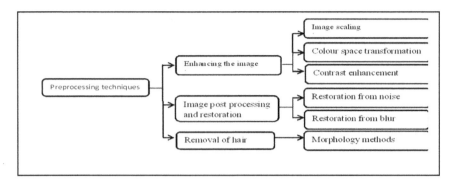

Fig. 1 The framework of pre-processing technique for the detection of skin cancer

techniques in pre-processing for hair removal. In Fig. 1, the data on pre-processing methods are summarized.

The paper is structured accordingly. The whole plan of the pre-processing phases is discussed in Sect. 2. The various methods for enhancing the image, restoring the image and removing the hair are subsequently discussed in Sects. 2.1, 2.2 and 2.3 and the most efficient methods are also illustrated by the literature. Paragraph 3 is the document's conclusion.

2 Pre-processing Techniques for Skin Cancer Detection System

Image processing is a necessary step for detecting skin cancer and in removing noise and improving image quality. The picture needs to be processed before precise identification in the design and evaluation of a computer-aided diagnosis scheme. It is the first and basic step in a further procedure that has a direct impact. The system's precision can be significantly improved by a good choice of pre-processing techniques [3]. The methods for pre-processing are split into three phases: image enhancement, image post-processing and restoration as well as hair removal. Figure 1 illustrates the overall structure of these methods during the pre-processing phase.

2.1 Enhancing the Image

Enhancing the picture is the method of improving the image quality by manipulating the picture to better depict skin cancer detection [4]. This is a crucial step to ensure more accurate processing of an enhanced and transformed image. The improvement of the picture is divided into three classifications. They are image scaling, transforming a colour space and contrast enhancement of the picture.

2.1.1 Image Scaling

Image scaling means picture resize owing to an absence of the same and standard size. Images of skin cancer are collected from various sources and can vary in size. It is the first step in redimensioning the image so that it contains the pixels with a fixed width but the height is normal.

2.1.2 Colour Space Transformation

The conversion of the colour space plays a significant part in the identification of skin cancer, and dermoscopic pictures are generally taken with a dermoscopic digital camera. The area of colour covers red, green, blue (RGB), hue, saturation, value (HSV), CIE-LAB, hue, saturation, intensity (HSI) and CIE-XYZ.

The RGB colour code is often used for image processing and is representative of the three primary spectral colours red, green and blue. But the colour space of RGB is restricted in high-level processing to the use of other colour representations. Colours in RGB image processing are usually presented. The resulting RGB colour images are often transformed into a scalar image because of the ease and the computational efficiency of the scalar processing and other coloured images have been created for high-level processing.

HSV and HSI colour models interpret comparable colours, e.g. hue, saturation and intensity, in terms of colour or shade. They represent the average wavelength of the spectral colour, colour and lightness of the colour. CIE-LAB, whereas the CIE-XYZ colour system can produce each colour with positive tristimulus value is another colour model is for uniformity.

The pictures used for the identification of skin cancer structures must demonstrate highly intensive colour differences; for the correct identification of the corners, the LAB colour space frequently depicted as 'LAB', as it contains the whole variety of colours, should be ideal. The 'L' is the axis where red and green colours, while the 'b' axis represents the yellow/blue opponent colours. The 'L' is the axis along which the colours are marked. This applies LAB as an intermediate colour space for picture conversion from RGB with XYZ. The primary picture is transformed into a greyscale picture that gives lightness values. As RGB colour space has certain constraints in high-level processing, there are alternative depictions of colour space. LAB is one of the useful colour models which, with the light components red/green and blue/yellow, represent every colour. The use of XYZ as an intermediary colour space could be helpful to transform the RGB into a LAB. The light would show the picture of the greyscale skin. Figure 2 shows the resulting sample of image enhancement on skin cancer images.

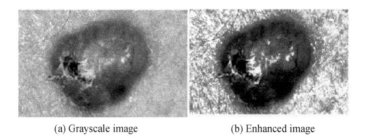

(a) Grayscale image   (b) Enhanced image

Fig. 2 Enhancing the image from greyscale to high contrast image

2.2 Contrast Enhancement

Improving the contrast is a helpful step in further processing and is a significant part of improving the picture quality. This enhanced image with simple restored algorithms enables border detection. It sharpens the border of the image and enhances image accuracy. Contrast improvement is categorized as an improvement in linear contrast and an increase in non-linear contrast [5].

2.2.1 Enhancement of Linear Contrast

Linear enhancement of contrast refers to stretching of contrast techniques. The greyscale values and the histogram can be stretched over the full distance to convert the image into higher contrast image. Linear contrast enhancement methods are classified into three categories; they are min–max linear contrast enhancement, percentage linear contrast enhancement stretch and piecewise linear contrast stretch as illustrated in Fig. 3.

The original min and the max image values are assigned by the newly specified values, which take advantage of the entire range of available image brightness. Take a picture of a min luminosity value of 45 and a max value of 205. If a picture of this type is regarded without improvement, it does not display the values 0 through 44 and 206–255. The min value from 45 to 0 and the max value from 120 to 255 can be detected to detect significant spectral differences. Linear contrast is same as the minimum linear contrast, except for certain min and max pixel percentage values of the mean histogram. This method uses the min values and max values that are specified.

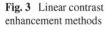

Fig. 3 Linear contrast enhancement methods

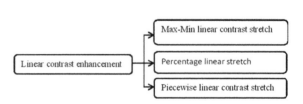

In a bi-or remodelling image, an analysis can stretch some histogram values to increase enhancement in selected areas. This method of improving contrast is known as piece by piece. A partially linear improvement of contrast involves the identification of several linear steps that increase the range of brightness in histogram mode.

2.2.2 Non-linear Enhancement of Contrast

Non-linear enhancement of contrast includes histogram equalization, adaptive histogram equalization, homomorphic filtering and unsharp masking as illustrated in Fig. 4.

- Histogram equalization: There are two techniques of histogram equalization, one of which is worldwide histogram equalization and one of the other of local histogram equalization. A small portion of an image is called global histogram equalization in local histogram technology. The histogram equalization strategy frequency values are mapped to certain intensity levels. In order to improve image contrast, the image histogram parallel approach is used. Histogram equalization is only inadequate to improve the quality of the medical image in low contrast because the luminosity of high contrast images can be decreased. For medical images, data conservation and excellent contrast improvement are very important to the precise diagnosis, both histogram equalization and adaptive histogram equalization. In order to accomplish this adaptive histogram equalization method, medical pictures were improved further.
- Adaptive equalization histogram: Adaptive equalization histogram correction plays a role in further improvement of medical images after equalization histogram. But medical images have little contrast and complicated background structures. Due to certain visual areas, there are more noise artefacts after improved contrast that might lead to misdiagnosis.
- Homomorphic filters: Homomorphic filtering has been used to resolve the limitations of adaptive histogram equalization to reduce noise in areas of low visual variation in medical images. This functions in the frequency domain and helps to correct the uniform illumination error with clinical images with low contrast.
- Unsharp masking: This is another technique to increase the sharpness and contrast of the image.

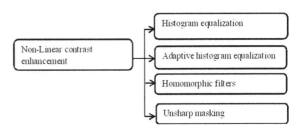

Fig. 4 Non-linear contrast enhancement methods

2.3 Image Restoration and Post-Processing

Image restore is the method for recovering the broken and rumbled image. It increases the picture appearance and converts the degraded picture into a noiseless and debilitated picture during restore. Various faults, such as picture system imperfection and poor concentrate, can cause picture degradation. The noise is divided into Gaussian, Salt and Pepper, Poison and Speckle in four groups [6]. Different picture restoration techniques like inverse filters, Weiner filters, less square filters, etc. are accessible.

2.3.1 Noise Restore

Many de-noise methods are available. The fundamental techniques can be categories such as spatial filtering and domain space transformation filters like mean, median, Wiener filters, etc. The description in the following [7] is more prevalent spatial filters to remove noises and smooth the picture region.

Mean filter:
Mean filters are of four types.

- Arithmetic mean filter: It is the simple type of the mid-filter also known as the linear filter. The noise can be even and works fine with Gaussian noise. This filter enables the variety in a picture to smoothing and blurs the picture.
- Geometric mean filter: It is identical with the mean arithmetical filter, but the medium filter can maintain the data on the picture better than the medium.
- Harmonic mean filter: It is the finest noise and salt filter for Gaussian noises but does not work for pepper noise.
- Contra-harmonic mean filter: This filter is the easiest way to eliminate salt and pepper noise. The edge can be preserved and the noise removed much superior to the medium filter.

Adaptive filters:

- Adaptive noise reduction filter: This filter is used for different noise removal for the image.

Order statistics filters:

- Median filter: This filter is a salt and pepper noise efficient filter without decreasing the picture sharpness.

Max filters and min filters:
The most bright and darkest points in the image are found through these filters. The min filter is the shortest point in the pixel and the max filter is the brightest point in the pixel filter. Max's filter enables you to locate pixels in a picture, while min's filter helps locate dark pixels.

- Mid-point filter: This filter is good for randomly distributed noises such as speckle noise.

A Study on Pre-processing Techniques ... 151

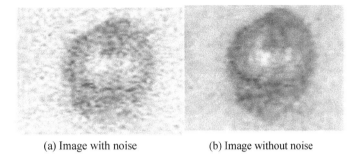

(a) Image with noise   (b) Image without noise

Fig. 5 Noise removed image

– Gaussian smoothing filter: This filter is useful for the image smoothing and sharpening.

Researchers use these most common filters for noise reduction in the preprocessing stage of detection systems to be median filters, adaptive medium filters, medium filters and Gaussian smoothing filters.

2.3.2 Blur Restore

Blur is one type of picture degrading and a method for incomplete image formation. It takes place through a bad focus from the original picture to the camera. There are various current de-blurring methods, such as Lucy Richardson algorithm methods, inverse filtering, Wiener de-blurring technology and approach to neural network. The Wiener filter was the most common technique in medical applications to remove noise is shown as follows (Fig. 5).

2.4 Hair Removal

There is always a common obstacle for thick hair analysis of small skin lesions and also has difficulties in the segmentation procedure. In the analysis of skin cancer images, scientists used certain methods like mathematical morphology [8] for detecting curvilinear structures, an approach based on paintings, dull razor and transforming the top-hat automated software in combination with a bicubic interpolation process.

A widespread morphological closure using three structuring elements that model three lines of orientation is the best way to find dark hair on light skin. It releases hair images by basic image interpolation of the two-linear colours. These operations are used to obtain hair-free images. The transformation of the top hat is a method to

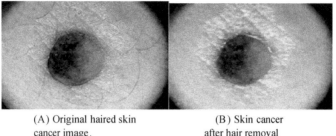

(A) Original haired skin cancer image.   (B) Skin cancer after hair removal

Fig. 6 Skin cancer images showing hair removal

remove small details and elements from the image. In the different image processing tasks, top-hat transformations are used; for example, feature retrieval, context equalization and improving images are shown in Fig. 6.

The resulting images can be seen in the pre-processed stage of the method for the identification of skin cancer and are nearly ready for the next step, the segmentation phase.

3 Conclusion

This article examines the methods for preparing the automatic detection system for skin cancer. It provides the expertise needed to help scientists assess the significance of high-level pre-processing methods, which require more effort to diagnose melanoma accurately. Three-section image enhancing and restoration of the image and hair removal are classifying the entire process. These methods assist to improve pictures of the skin cancer and some of the filters are used in noise removal and the pictures are smoothened.

References

1. M. Silveira, J.C. Nascimento, J.S. Marques, A.R. Marçal, T. Mendonça, S. Yamauchi, J. Maeda, J. Rozeira, Comparison of segmentation methods for melanoma diagnosis in dermoscopy images. IEEE J. Sel. Top. Sign. Proces. **3**(1) (Feb 2009)
2. O.V. Michailovich, A. Tannenbaum, Despeckling of medical ultrasound images. IEEE Trans. Ultrason. Ferroelectr. Freq. Control
3. H.T. Lau, A. Al-Jumaily, Automatically early detection of skin cancer: Study based on neural network classification. International conference of soft computing and pattern recognition, 2009
4. R. Garnavi, M. Aldeen, M.E. Celebi, A. Bhuiyan, C. Dolianitis, G. Varigos, Automatic segmentation of dermoscopy images using histogram thresholding on optimal color channels. Int. J. Biol. Life Sci. (2012)
5. S.S. Al-amri, N.V. Kalyankar, S.D. Khamitkar, Linear and non-linear contrast enhancement image. IJCSNS Int. J. Comput. Sci. Netw. Secur. **10**(2), 139–145 (Feb 2010)

6. P. Patida, M. Gupta, S. Srivastava, A. K. Nagawa. Image de-noising by various filters for different noise. Int. J. Comput. Appl. **9**(4) (Nov 2010)
7. B. Shinde, D. Mhaske, M. Patare, A.R. Dani, Apply different filtering techniques to remove the speckle noise using medical images. Int. J. Eng. Res. Appl. (IJERA), **2**(1), (Jan–Feb 2012)
8. P. Schmid, Segmentation of digitized dermatoscopic images by two-dimensional color clustering. IEEE Trans. Med. Imaging, (1999)

Prediction of Cricket Players Performance Using Machine Learning

P. Aleemulla Khan, N. Thirupathi Rao and Debnath Bhattacharyya

1 Introduction

In the current article, player's performance can be predicted by analyzing their past statistics and characteristics. Cricket player's abilities and performance can be measured in terms of different stats [1, 2]. Batsmen's statistics include batting average, batting strike rate, number of centuries, etc. Whereas bowlers' statistics are measured by bowling average, bowling strike rate, economy rate, etc, other characteristics of batsmen include batting hand of the batsman, the position at which the batsman bats, etc. and those of bowlers include, the type of bowler, bowling hand of the bowler, etc.

Moreover, recent performances of the batsman/bowler, the performance of the batsman/bowler against particular team and the performance of the batsman/bowler at a given venue are also taken into account for predicting his performance in the upcoming match [3, 4]. The team management, the coach and the captain utilize these facts and their own experience to select the team for a given match. In this paper, we used machine learning and data mining techniques to predict batsmen and bowler's performances in a given day's match. We experimented with four supervised machine learning algorithms and compared their performance [5]. The models generated by these algorithms can be used to predict the player's performance in future matches.

P. Aleemulla Khan (✉) · N. Thirupathi Rao · D. Bhattacharyya
Department of Computer Science and Engineering, Vignan's Institute of Information Technology (Autonomous), Visakhapatnam, AP, India
e-mail: p_aleekhan@yahoo.in

N. Thirupathi Rao
e-mail: nakkathiru@gmail.com

D. Bhattacharyya
e-mail: debnathb@gmail.com

© Springer Nature Singapore Pte Ltd. 2020
J. Fiaidhi et al. (eds.), *Smart Technologies in Data Science and Communication*, Lecture Notes in Networks and Systems 105, https://doi.org/10.1007/978-981-15-2407-3_20

2 Proposed Work

- The main purpose of this project is to predict a match that it will win or not. This project is mainly used for prediction of winning or losing the match.
- The main purpose of the project is for betting. We perform extensive simulation studies on an undergraduate cricket player dataset collected over 10 years across 11 different player's data from cricket information Web site.

3 Working Mechanism

The below procedure illustrates the prediction of the performance of the player using machine learning. The mechanism is as follows:

Step 1: Gathering data methods such as representation of quality of data.
Step 2: Remove the noisy data and correct inconsistencies in data.
Step 3: Directed acyclic graph (DAG) is used to represent different subsets of courses.
Step 4: Training base predictors are used to represent logistic regression.

The below diagrams show that workflow of the proposed system for the prediction of the performance of the player.

4 Results and Analysis

The below information shows the representation of accuracy 1 and accuracy 2 of cricket player's details (Figs. 1 and 2).

The below code shows the representation of prediction of player's performance that is runs scored, balls faced, wickets taken, etc. (Figs. 3, 4, 5, 6, 7 and 8).

Test result: Fail
Since both the input teams are same, there will be an error (Fig. 9).
Test Result: Pass
Since both the input teams are different, so there is no error.

5 Conclusion

In the current paper, a cricket selection model of players for the best cricket team was done by using machine learning models with the usage of various bowlers and batsmen datasets. Various performance metrics were considered to analyze the performance of the players like the weather conditions, previous scores, previous number of wickets, number of maiden overs, etc. Several machine learning algorithms

```
('Team Name1: ', 'Australia')
('Team Name2: ', 'India')
('Date till which to scrap: ', '20+Sep+2017')
('Ground ID: ', u'292')
Moving to the next Window...
[[u'KR Patterson', u'505110', 'Batsman'], [u'NM Lyon', u'272279', 'Batsman'],
'JR Hazlewood', u'288284', 'Batsman'], [u'TM Head', u'530011', 'Batsman'], [u'B
Stanlake', u'533042', 'Batsman'], [u'AJ Finch', u'5334', 'Allrounder'], [u'CA L
nn', u'326637', 'Allrounder'], [u'NM Coulter-Nile', u'261354', 'Allrounder'],
'A Zampa', u'379504', 'Bowler'], [u'JA Burns', u'326632', 'Bowler'], [u'AT Care
', u'326434', 'Bowler']]
()
[[u'KK Ahmed', u'942645', 'Batsman'], [u'KH Pandya', u'471342', 'Batsman'], [u'
S Dhoni', u'28081', 'Batsman'], [u'I Sharma', u'236779', 'Batsman'], [u'M Marka
de', u'1081442', 'Batsman'], [u'AM Rahane', u'277916', 'Allrounder'], [u'GH Vih
ri', u'452044', 'Allrounder'], [u'Shubman Gill', u'1070173', 'Allrounder'], [u'
A Jadeja', u'234675', 'Bowler'], [u'S Dhawan', u'28235', 'Bowler'], [u'V Shanka
', u'477021', 'Bowler']]
 _ 0 ... _ 1 ... _ 2 ... _ 3 ... _ 4 ... _ 5 ... _ 6 ... _ 7 ... _ 8 ...
9 ... _ 10 ... _ 0 ... _ 1 ... _ 2 ... _ 3 ... _ 4 ... _ 5 ... _ 6 ... _ _
 ... _ 8 ... _ 9 ... _ 10 ... bat_ground_1:
       Player Name Player id Number of matches ... 150s 200s Total Team Score
0      KR Patterson    505110                 0 ...    0    0                0
1           NM Lyon    272279                 0 ...    0    0                0
2      JR Hazlewood    288284                 0 ...    0    0                0
3           TM Head    530011                 0 ...    0    0                0
4         B Stanlake   533042                 0 ...    0    0                0
5          AJ Finch      5334                 0 ...    0    0                0
6           CA Lynn    326637                 0 ...    0    0                0
7    NM Coulter-Nile   261354                 0 ...    0    0                0
```

Fig. 1 Accuracy 1

```
[8 rows x 11 columns]
ball_ground_1:
       Player Name Player id Number of matches ... Wickets Taken 3Ws 5Ws
0          AJ Finch      5334                 0 ...             0   0   0
1           CA Lynn    326637                 0 ...             0   0   0
2    NM Coulter-Nile   261354                 0 ...             0   0   0
3           A Zampa    379504                 0 ...             0   0   0
4          JA Burns    326632                 0 ...             0   0   0
5          AT Carey    326434                 0 ...             0   0   0

[6 rows x 9 columns]
bat_form_1:
       Player Name Player id Number of matches ... 150s 200s      Total Team S
core
0      KR Patterson    505110                 0 ...    0    0
  0
1           NM Lyon    272279                 6 ...    0    0      [][][][]
[][]
2      JR Hazlewood    288284                 9 ...    0    0    [][][][][][][]
[][]
3           TM Head    530011                10 ...    0    0    [][][][][][][]
[][]
4         B Stanlake   533042                 1 ...    0    0
 []
5          AJ Finch      5334                10 ...    0    0    [][][][][][][]
[][]
6           CA Lynn    326637                 1 ...    0    0
 []
7    NM Coulter-Nile   261354                10 ...    0    0    [][][][][][][]
[][]
```

Fig. 2 Accuracy 2

```python
import pandas as pd


def oposition_score(df):
    df_score = 0
    for i in range(len(df)):
        if(df.loc[i,"BowlersFaced"] != 0):
            SR = (df.loc[i,"Runs"]/df.loc[i,"BallsFaced"])*100
            if(df.loc[i,"Wickets"] != 0):
                runs_per_wkt = df.loc[i,"Runs"]/df.loc[i,"Wickets"]
            else:
                runs_per_wkt = df.loc[i,"Runs"]
            additional = df.loc[i,"4s"] + 1.5*df.loc[i,"6s"]
            wkts = df.loc[i,"Wickets"]*(-5)


            score = SR + runs_per_wkt + additional + wkts
            df_score += score


    return df_score


def ball_score(df):
    df_score = 0
    for i in range(len(df)):
        if(df.loc[i,"Number of matches"] != 0):
            hauls = (df.loc[i,"3Ws"]*37.5 + df.loc[i,"5Ws"]*62.5)/df.loc[i,"Number of matches"]
```

Fig. 3 Code for prediction of player's performance

```
        wkts = df.loc[i,"Wickets Taken"]/df.loc[i,"Number of matches"]*100

        maidens = df.loc[i,"Maidens"]/int(df.loc[i,"Balls"]/6)*100*5

        overs = int(df.loc[i,"Balls"]/6) + float((int(df.loc[i,"Balls"])%6))/10

        economy = (1-((df.loc[i,"Runs given"])/(10*overs)))*100


        score = hauls + wkts + maidens + economy


        df_score += score

    return df_score


def bat_score(df):

    df_score = 0

    for i in range(len(df)):

        if(df.loc[i,"Number of matches"] != 0):

            SR = (float(df.loc[i,"Runs"])/float(df.loc[i,"Balls Faced"])*100)

            runs = (float(df.loc[i,"Runs"])/float((df.loc[i,"Total Team Score"])/10))*100

            wkts = (1-((float(df.loc[i,"Number of Dismissals"]))/(float(df.loc[i,"Number of matches"]))))*100

            score50 = (float(df.loc[i,"50s"])/float(df.loc[i,"Number of matches"]))*50

            score100 = ((float(df.loc[i,"100s"])/float(df.loc[i,"Number of matches"]))*100)

            score150 = ((float(df.loc[i,"150s"])/float(df.loc[i,"Number of matches"]))*150)

            score200 = ((float(df.loc[i,"200s"])/float(df.loc[i,"Number of matches"]))*200)


            score = SR + runs + wkts + score50 + score100 + score150 + score200


            df_score += score

    return df_score
```

Fig. 3 (continued)

Fig. 4 Initial details

Fig. 5 List of teams

were run on these datasets and identified random forest model will give us the best results compared to other algorithms for our constraints.

Fig. 6 List of players

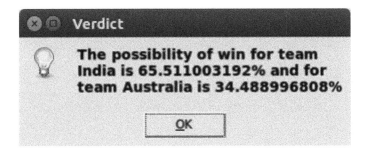

Fig. 7 Predicted result

Variable	Input Values
Team 1	India
Team 2	India

Fig. 8 Test case 1

Variable	Input Values
Team 1	India
Team 2	Australia

Fig. 9 Test case 2

References

1. S. Muthuswamy, S.S. Lam, Bowler performance prediction for one-day international cricket using neural networks, in *Industrial Engineering Research Conference* (2008)
2. I.P. Wickramasinghe, Predicting the performance of batsmen in test cricket. J. Hum. Sport Exerc. **9**(4), 744–751 (2014)
3. G.D.I. Barr, B.S. Kantor, A criterion for comparing and selecting batsmen in limited over's cricket. Oper. Res. Soc. **55**(12), 1266–1274 (2004)
4. S.R. Iyerand, R. Sharda, Prediction of athletes performance using neural networks: an application in cricket team selection. Expert Syst. Appl. **36**, 5510–5522 (2009)
5. D. Bhattacharjee, D.G. Pahinkar, Analysis of performance of bowlers using combined bowling rate. Int. J. Sports Sci. Eng. **6**(3), 1750–9823 (2012)

Using *K*-means Clustering Algorithm with Python Programming for Predicting Breast Cancer

Prasanna Priya Golagani, Shaik Khasim Beebi and Tummala Sita Mahalakshmi

1 Introduction

Cancer causes severe metabolic changes in the cell. In cancer, the cells do not die when they have to die, and new cells are born when they are not required. Cancer cell divides uncontrollably and produces numerous new cells. Cancer is of two types benign and malignant. Malignant tumors are cancerous and can invade the surrounding tissues, benign tumors do not spread to other tissues they are local to their site and sometimes they can be quite large. Cancer Statistics: Cancer stands second worldwide to cause death. During 2015, 8.75 million humans passed away out of malignant growth. In 2017, an estimated no. of new cancer cases are 1,688,780. Nearly 600,920 people are estimated to die with cancer in 2017. For every 8 min, one woman dies with cervical cancer. In India, an estimated number of 2.5 million people are suffering from cancer. Every year an estimated number of 7 lakh cases are registered in India. Nearly 556,400 people are dying with cancer every year in India. Based on the primary site of origin cancer can be divided into different types like (1) breast malignant growth, (2) lung malignant growth, (3) prostrate malignant growth, (4) liver malignant growth, (5) renal cell carcinoma, (6) oral cancer and (7) brain cancer.

Breast Cancer: A malignant growth in the chest is called breast carcinoma. Breast cancer can be classified based on histopathology, stage (TNM), category and receptor position in addition to the existence or non-existence of genetic code in the DNA.
Breast Cancer Statistics: There is an estimation that in 2017, 252,710 recently developed occurrences of invasive breast carcinoma will be detected in females and 2470 in males. Along with it, 63,410 recently developed occurrences of in situ breast

---

P. P. Golagani (✉) · T. S. Mahalakshmi
Department of Computer Science and Engineering, GITAM University, VSP, Visakhapatnam, AP, India
e-mail: prasannapriya18@gmail.com

S. K. Beebi
Department of Biotechnology, GITAM University, VSP, Visakhapatnam, AP, India

© Springer Nature Singapore Pte Ltd. 2020
J. Fiaidhi et al. (eds.), *Smart Technologies in Data Science and Communication*, Lecture Notes in Networks and Systems 105, https://doi.org/10.1007/978-981-15-2407-3_21

cancer will be detected in females. Nearly 40,610 female and 460 male deaths from breast cancer are estimated in 2017. Advanced rate of 3.45 million women was alive with a chronology of breast carcinoma on January 01, 2017 [1].

1.1 Signal Transduction Pathways

In the breast cancerous cell, the cell signaling is affected. Apoptosis in inhibited and new cells are produced continuously, and this is due to the changes in some of the signal transduction pathways. In this paper, we studied some of these signal transduction pathways like PKB, MAPK, MTOR, FasL, Notch, SHH, Tnf and Wnt.

1.2 Computational Modeling of the Pathways Using SBML and Cell Designer

Converting the biological pathway into a computable model helps in analyzing it rapidly using simulation and other mathematical methods. We used SBML for computational modeling of signal transduction pathways. The programs written in SBML for each pathway are executed in Cell Designer.

1.3 Data Mining of Simulated Pathways

The simulated pathways are stored as xml files. The data from the xml files is extracted into tables and subjected to data preprocessing techniques like data cleaning, data integration, data transformation, data reduction and data discretization.

1.4 Data Clustering

A data set obtained after data preprocessing is used for clustering. *K*-means clustering algorithm is used to obtain clusters. *K*-means clustering algorithm is implemented using Python program.

2 Literature Review

Tanya and Aditya described a 17 genetic code binary impression of chest broadcasting malignant cell deduced transcripts enhanced from blood, enabling high responsiveness primal tracking of backlash [2]. Baowen Yuan and Simon Schafferer recognized metabolites that have dormant value for evolution of a multimarker blood-based test to complement and develop early breast cancer identification [3]. Wei-Bing Yin and Ming-Guang Yan investigated the circular RNA revelation outline in chest carcinoma fringe blood and tried to uncover symptomatic biomarkers [4].

Angel Cruz-Rao and Hannah Gilmore presented a traditional method called high-throughput adjustable sampling for entire slide histopathology image scrutiny involving an adjustable sampling procedure ground on chances gradient and quasi-Monte Carlo sampling and a portrayal studying categorizer ground on conventional neural networks [5]. Jun Xu and Lei Xiang used a stacked sparse auto-encoder as an occurrence of deep studying for productive nuclei spotting on high-resolution histopathological images of chest carcinoma [6].

Babak and Mitko assessed the execution of deep studying algorithms for pathological diagnosis of breast cancer against detecting the metastases in tissue sections of women with chest carcinoma [7]. Mitko and Paul proposed an assessment of mitosis detection algorithm for counting the mitotic figures in hematoxylin and eosin-stained histology sections [8]. Jose and Alyssa reviewed advantages of dynamic infrared thermography, short comings and opportunities of it in the future development [9].

3 Proposed Work

Breast cancerous cell has the changes in signal transduction pathways when compared to a normal breast cell. In this paper, we are studying some of these pathways like PKB, MAPK, MTOR, Fas ligand, Notch, SHH, Tnf and Wnt pathways.

Converting these biological pathways into a computable model helps in analyzing it rapidly. We are using SBML for computational modeling of signal transduction pathways. The programs written in SBML for each pathway are executed in Cell Designer.

These simulated pathways are stored in the form of xml files. We extracted the data in the xml files into tables and we applied information processing techniques to it like information cleaning, information integration, information transformation, information reduction and information discretization. Thus, obtained data set is subjected to clustering by implementing *K*-means clustering algorithm in python language to form two clusters one representing benign and the other representing malignant breast cancer.

4 Architectural Diagram of Our Work

Figure 1 shows the flowchart of the stepwise procedure involved in the work.

5 Results and Discussions

Tanya and Aditya described a 17 genetic code binary impression of chest broadcasting malignant cell deduced transcripts enhanced from blood, enabling high responsiveness primal tracking of backlash [2]. They just specified the genes of a blood cell, they did not study any of the cell physiology nor did they use any of the tools for simulating any of the physiology. Whereas in this paper apart from specifying some of the computational tools that can be used we also studied the breast cancerous cell physiology in depth and identified some physiological aspects that can be modeled using computational tools, and we identified two tools that can be used that are SBML and Cell Designer and we used them effectively. Baowen Yuan and Simon Schafferer recognized metabolites that have dormant value for evolution of a multimarker

Fig. 1 Architectural diagram of our work

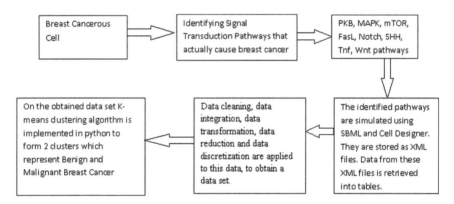

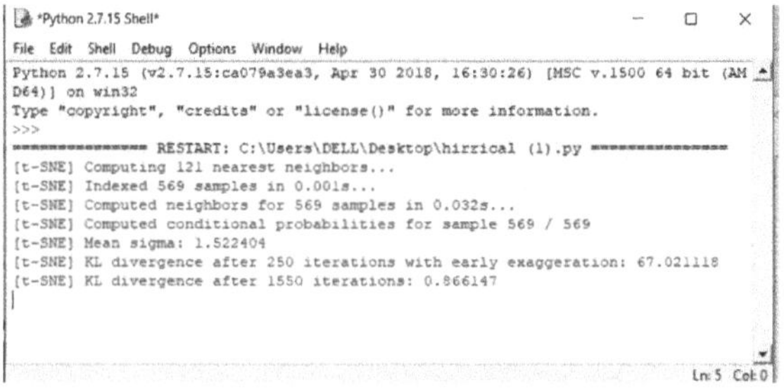

Fig. 2 Running the Python program

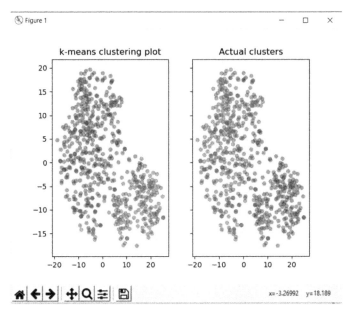

Fig. 3 *K*-means clustering plot

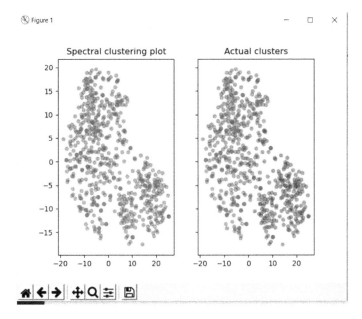

Fig. 4 Spectral clustering plot

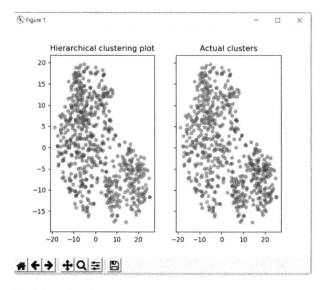

Fig. 5 Hierarchical clustering plot

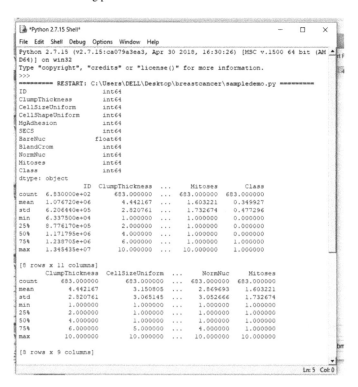

Fig. 6 Running the shell program

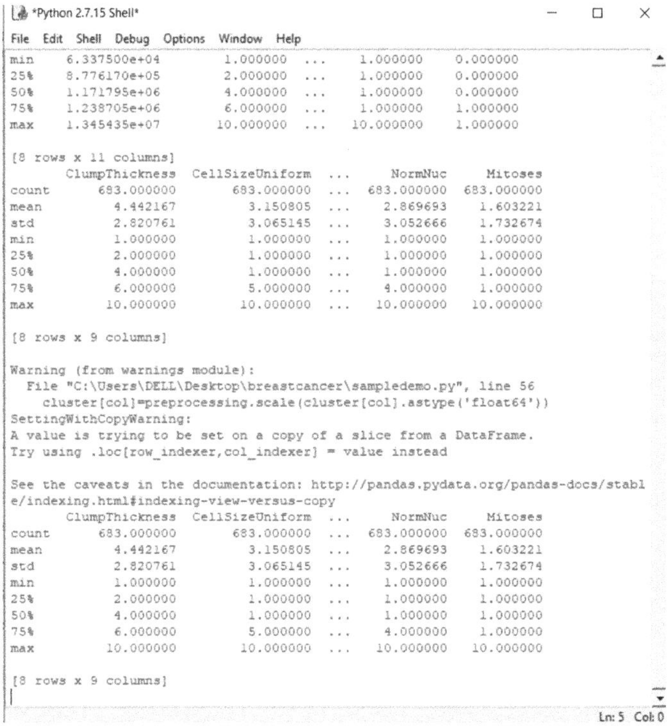

Using K-means Clustering Algorithm with Python … 169

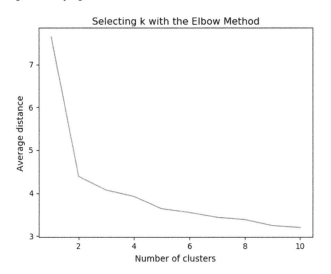

Fig. 7 Running the shell program

Fig. 8 Graph generated after running the Python code

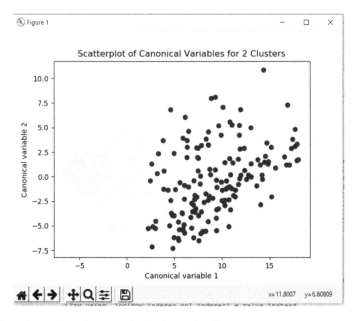

Fig. 9 Two clusters formed representing benign and malignant cancer

blood-based test to complement and develop early breast cancer identification [3]. Yuan and Simon identified the metabolites that may help in cancer detection, but they did not specify the deviations caused in them due to cancer. Whereas in this paper we are not only doing computational modeling of the pathways, but we also specified the deviations in the pathways and explained how they are becoming the root cause for cancer. Wei-Bing Yin and Ming-Guang Yan investigated the circular RNA revelation outline in chest carcinoma fringe blood and tried to uncover symptomatic biomarkers [4]. Wei-Bing Yin and Yan concentrated only on the RNA expression profiles involved in the pathway. In our paper, we are studying the deviations of the pathway in a breast cancerous cell and identifying how it can be modified. In our paper, we modeled totally eight pathways. Figure 2 shows the screen shot of running the code in python. Figures 3, 4 and 5 shows the cluster plots for K-means, spectral and hierarchical clustering. Figures 6 and 7 shows the screenshots of running the shell program. Figure 8 shows the graph of selecting the K value by Elbow method. Figure 9 shows the resultant clusters showing benign and malignant tumors.

6 Conclusion

The complete study of the signal transduction pathways which are mutated because of breast cancer helps researchers to study the morphology of the breast cancerous cell better and they can start to work to change the morphology of the breast cancerous cell into a normal breast cell, in spite of starting from the beginning of how the pathway actually works and what is the deviation. Computational modeling of these pathways helps in making this data readily available to them in a format where the exact locations where the pathways have deviated are specified.

References

1. K.D. Miller, R.L. Siegel, C.C. Lin et al., Cancer treatment and survivorship statistics. CA Cancer J. Clin. **66**, 271–289 (2016)
2. T.T. Kwan, A digital RNA signature of circulating tumor cells predicting early therapeutic response in localized and metastatic breast cancer. AACR J (2018). https://doi.org/10.1158/2159-8290.cd-18-0432
3. B. Yuan, S. Schafferer, Q. Tong, A plasma metabolite panel as biomarkers for early primary breast cancer detection. Int. J. Cancer **13** (2018)
4. W.-B. Yin, M.-G. Yan et al., Circulating circular RNA hsa_circ_0001785 acts as a diagnostic biomarker for breast cancer detection. Clinica Chimica Acta **487**, 363–368 (2018)
5. A. Cruz-Roa, H. Gilmore, High-throughput adaptive sampling for whole-slide histopathology image analysis (HASHI) via Convolutional neural networks: application to invasive breast cancer detection. Research Gate, 24 May (2018)
6. J. Xu, L. Xiang, Stacked sparse autoencoder (SSAE) for nuclei detection on breast cancer histopathology images. IEEE Trans. Med. Imag. **35**(1), 119–130 (2016)
7. P.J. van Diest, B. van Ginneken, Diagnostic assessment of deep learning algorithms for detection of lymph node metastases in women with breast cancer. Med. Image Anal. **318**(22), 2199–2210 (2017)
8. M. Veta, P.J. van Diest, Assessment of algorithms for mitosis detection in breast cancer histopathology images. Med. Image Anal. **20**(1), 237–248 (2015)
9. J.L. Gonzalez-Hernandez, et al., Technology, application and potential of dynamic breast thermography for the detection of breast cancer. Int. J. Heat Mass Transf. **131**, 558–573 (2019)

Compact Slot-Based Mimo Antenna for 5G Communication Application

Sourav Roy, Srinivasa Naik, S. Aruna and S. K. Gousia Begam

1 Introduction

In this busy world, everyone looking for a quick response and they want to complete their needs quickly and correctly from long distances. To achieve higher data rates, long-distance communication, increased channel capacity and reduced bit error rate (BER), 5G MIMO technology came to fulfill all these requirements [1].

For enhancing bandwidth and gain of the antenna, DGS is used [2]. Spiral head, H-shaped slot and open-loop dumbbell are the different DGS structures [3]. Due to disturbance in ground surface will affect the current distribution which causes changes in capacitances and inductance of the transmission for better performance [4, 5]. We can use multiple DGS structures to increase bandwidth.

In [6], a compact MIMO antenna with size $43.5 \times 43.5 \times 1.6$ mm^3 is designed for UWB application. And the frequency range of the antenna is from 3.1 to 10.6 GHz. For better isolation, a fence-like strip is placed diagonally. A four-element MIMO antenna for 5G applications at 28 GHz is presented in [7], it covers a bandwidth of 1.78 GHz from 27.10 to 28.88 GHz and the gain is 14.8 dBi. In [8], a two-element (2×1) MIMO antenna is presented; it is designed for dual-band operation at 28 and 37 GHz by placing an L-shaped slot on the patch. The antenna gives excellent return loss and gain, but the bandwidth is only 1.02 GHz.

We designed a MIMO antenna of compact size in this paper at 28 GHz frequency using Rogers RT/duroid 5880 material and L-shaped slot at the right corner of the patch to attain good performance. The proposed MIMO antenna is designed, simulated and analyzed using 3D electromagnetic solver ANSYS HFSS [9].

S. Roy (✉) · S. Naik
Vignan's Institute of Information Technology, Visakhapatnam, India
e-mail: sourav31roy@gmail.com

S. Aruna · S. K. Gousia Begam
Andhra University College of Engineering (A), Visakhapatnam, India

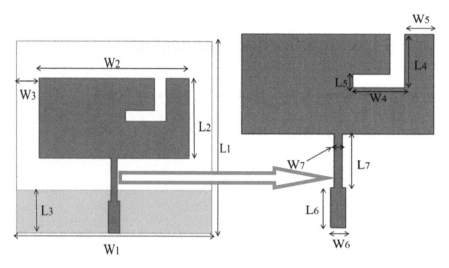

Fig. 1 Design geometry of single antenna

2 Single-Antenna Design

The antenna is designed on the Rogers RT/duroid 5880 material as substrate with h = 1.57 mm thickness, loss tangent δ = 0.0009 and εs_r (dielectric constant) = 2.2. The design geometry of the single antenna is shown in Fig. 1. There is a L-shape cut in the patch, and partial ground is used for desired performance.

The overall dimension of the antenna is 17 × 18 mm^2. The design parameters of the antenna are W_1 = 17 mm, L_1 = 18 mm, W_2 = 13 mm, L_2 = 7.5 mm, W_3 = 2 mm, L_3 = 4 mm, W_4 = 3.5 mm, L_4 = 4 mm, W_5 = 2 mm, L_5 = 1 mm, W_6 = 1 mm, L_6 = 3 mm, W_7 = 0.5 mm, L_7 = 4 mm. Figure 2 depicts there turn loss of the single antenna. The antenna covers the frequency band from 26.19 to 29.48 GHz. Figure 3 illustrates the 3D polar gain plot of the single antenna.

3 Two-Element MIMO Antenna Design and Analysis

Figure 4 illustrates the two-element MIMO antenna. The overall dimension of the two-element MIMO antenna is 31.1 × 18 mm^2. The separation between the two antennas is 1.1 mm. Figure 5 depicts the S-parameters of the antenna. This antenna covers 26.61–29.27 GHz frequency bands. The S_{12} and S_{21} of this antenna are nearly same.

The ECC of the MIMO antenna is calculated using radiation pattern method, and the formula is given below (Eq. 1)

Compact Slot-Based Mimo Antenna ... 175

Fig. 2 S_{11}(dB) versus frequency plot of the single antenna

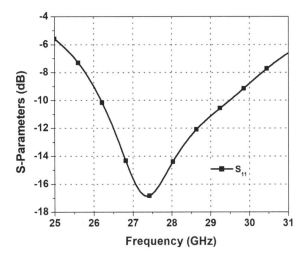

Fig. 3 Three-dimensional polar plot of single antenna at 28 GHz

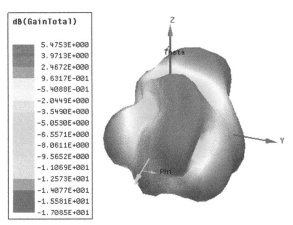

Fig. 4 Design geometry of two-element MIMO

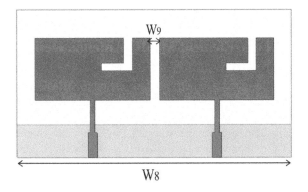

Fig. 5 S-parameters of the MIMO antenna

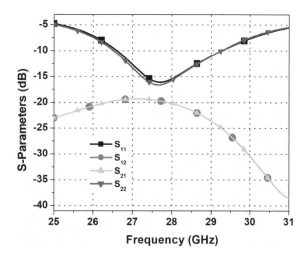

$$\rho_e = \frac{\left| \iint_{4\pi} [F_1(\theta, \emptyset) * F_2(\theta, \emptyset) d\Omega] \right|^2}{\iint_{4\pi} |F_1(\theta, \emptyset)|^2 d\Omega \iint_{4\pi} |F_2(\theta, \emptyset)|^2 d\Omega} \quad (1)$$

The D.G of the MIMO antenna is calculated using ECC. The mathematical expression for the calculation of D.G is given below (Eq. 2).

$$\text{D.G} = 10 \sqrt[10]{1 - (\text{ECC})^2} \quad (2)$$

The simulated ECC and D.G are >0.006 and >9.99 in the required frequency range considering port 1 and port 2. The D.G and ECC are very essential parameters for MIMO antenna. The simulated ECC and D.G parameters are depicted in Fig. 6.

4 Four-Element MIMO Antenna Design and Analysis

The designed geometry of four elements MIMO antenna is represented in Fig. 7. The size of antenna is 31.1×30.1 mm^2. The gap between antennas is $W_9 = L_9 = 1.1$ mm. The simulated S-parameters of the MIMO antenna are shown in Fig. 8a–d. This MIMO antenna covers 26.55–29.27 GHz frequency spectrum. The simulated ECC of the antenna 3 is shown in Fig. 9a–d with all combinations. In all combinations, the ECC is less than 0.5 which is very good for MIMO applications. The simulated D.G of the MIMO antenna with four elements is shown in Fig. 10a–d. In all cases, the D.G is found near about 9.98.

Fig. 6 a ECC and **b** D.G of the MIMO antenna by considering port 1 and port 2

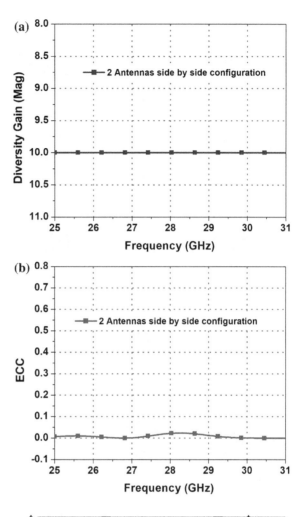

Fig. 7 Design geometry of MIMO antenna 3 (up-down configuration)

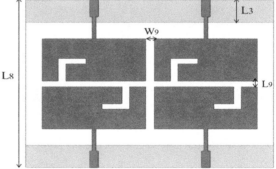

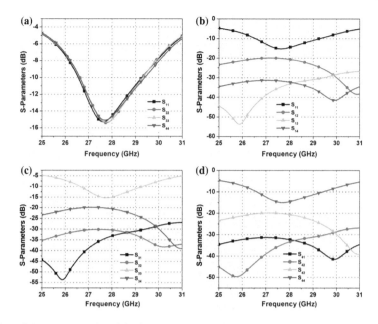

Fig. 8 a–d *S*-parameters plot of the four-element MIMO antenna

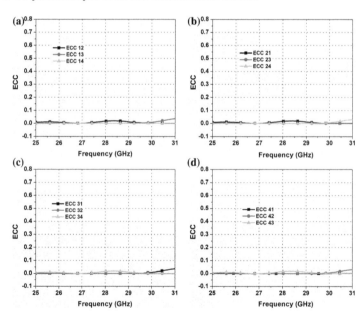

Fig. 9 a–d ECC parameters of the four element MIMO antenna

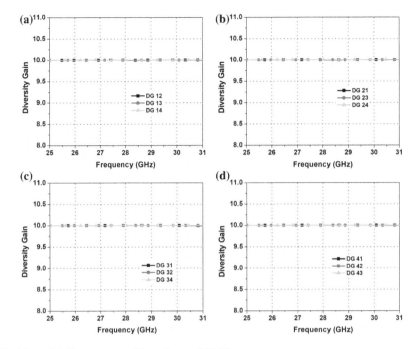

Fig. 10 a–d D.G parameters of four-element MIMO antenna

5 Conclusion

In this article, a compact two-element and four-element MIMO antennas are designed using Rogers RT/duroid 5880 substrate and analyzed using ANSYS HFSS software. The proposed four-element MIMO antenna covers a frequency band from 26.55 to 29.27 GHz, and the gain of the antenna found is 6.09 dB. The D.G of the antenna is >9, and ECC of the antenna is <0.035.

Acknowledgements This work is financially supported by DST-SERB with File No: EEQ/2016/000391.

References

1. M.G. Kachhavay, A.P. Thakare, 5G technology-evolution and revolution. Int. J. Comput. Sci. Mobile Comput. **3**(3), 1080–1087 (2014)
2. M.K. Khandelwal, B.K. Kanaujia, S. Kumar, Defected ground structure: fundamentals, analysis, and applications in modern wireless trends. Int. J. Antennas Propag. (2017)
3. J. Liu, W.Y. Yin, S. He, A new defected ground structure and its application for miniaturized switchable antenna. Prog. Electromagn. Res. **107**, 115–128 (2010)

4. L.H. Weng, Y.C. Guo, X.W. Shi, X.Q. Chen, An overview on defected ground structure. Prog. Electromagn. Res. **7**, 173–189 (2008)
5. G. Breed, An introduction to defected ground structures in microstrip circuits. High Freq. Electron. **7**, 50–54 (2008)
6. H. Qin, Y.F. Liu, Compact UWB MIMO antenna with ACS-fed structure. Prog. Electromagn. Res. **50**, 29–37 (2014)
7. Y. Rahayu, L. Afif, M.R. Radhelan, I. Yasri, F. Candra, Design of 28 GHz microstrip MIMO antennas for future 5G applications. Sinergi: Jurnal Teknik Mercu Buana **22**(3), 149–154 (2018)
8. A. Rachakonda, P. Bang, J. Mudiganti, A compact dual band MIMO PIFA for 5G applications, in Materials Science and Engineering Conference Series, vol. 263, No. 5, p. 052034, Nov 2017
9. HFSS ver.17. Ansoft Corporation, Pittsburgh, PA

DGS-Based Wideband Microstrip Antenna for UWB Applications

Y. Sukanya, Viyapu Umadevi, P. A. Nageswara Rao, Ashish Kumar and Rudra Pratap Das

1 Introduction

Presently, ultra-wideband antennas are having significant applications in communication area [1]. Because of simplicity of structure, large impedance bandwidth and wide ranging patterns monopole have become quite popular in the 3.1–10.6 GHz range [2, 3]. Portable devices such as microstrip antennas have less weight, size and cost [4–6]. These can be embedded in cellular phones, pagers as well as aircrafts and satellites.

The chief demerit of microstrip antenna reduced gain and small bandwidth. To circumvent this problem, DGS (defected ground structure) can be used. In this method, a defect is etched on the ground plane causing disturbance in the pattern of current and spreading of waves [7]. If the substrate permittivity exceeds unity, then the surface waves appear. So for high losses caused by surface waves, the microstrip antenna has some difficulties in gain, efficiency and bandwidth. The waves of the surface propagate and come to the edges can get reflected due to diffraction. In fabrication, back radiation can increase surface waves if dielectric constant is high.

For defected ground structure, many types of geometrical shapes can be used such as circles, squares, spirals and L- or H-shaped defects [8]. In some cases, rectangle or dumbbell shapes are preferred. Such antenna provides higher bandwidth and reduced loss on return. Dual-band, triple-band and multi-band characteristics also can be achieved with defected ground structures [9–13].

Y. Sukanya (✉) · V. Umadevi · Ashish Kumar
Department of ECE, Vignan Institute of Information Technology, Visakhapatnam, India
e-mail: sukanyagitam@gmail.com

P. A. Nageswara Rao
Department of ECE, G V P College of Degree and PG Courses, Visakhapatnam, India

R. P. Das
Department of ECE, NSRIT, Visakhapatnam, India

2 Parameters of Antenna

Table 1 shows the parameters of proposed antenna. It has radiating patch in the upper part. This comprises of semicircular radiating element and semicircle strip along with smiley. The size of the substrate is 29.5 mm × 25 mm. The ring patch has radii, outer radius R_1 and inner radius R_2. The semicircular strip has radii, outer radius R_3 and inner radius R_4 as shown in Fig. 1.

The first ring patch gets input from 50 Ω feed line having width of 2.4 mm and length of 12.06 mm. A part of ground portion is kept on the back ground having dumbbell-shaped defect. The separation between the ring and strip is 's,' which decides the impedance.

The antenna is simulated on FR4 material having dielectric constant of 4.4, thickness of 1.59 mm and dielectric loss tangent of 0.02. The bottom side the DGS method is applied with half ground plane having $L_g = 11.5$ mm. The dumbbell shape is etched on the ground plane. The parameters for dumbbell are dumbbell size (D_s

Table 1 Parameters of the designed antenna

Parameter	Size (mm)	Parameter	Size (mm)
L_s	29.5	L_g	11.5
W_s	25	D_b	0.2 * 1
L_f	12.06	D_s	0.7 * 0.5
W_f	2.4	D_g	1.3
R_1	12	H	1.59
R_2	9	S	1.5
R_3	8	D	2.51
R_4	5.5		

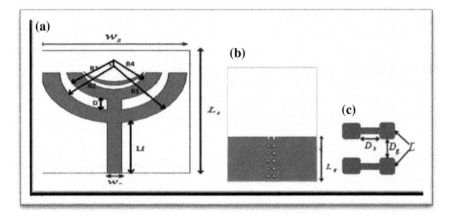

Fig. 1 Geometry of proposed antenna. **a** Front view, **b** back view and **c** inner dumbbell dimensions

Fig. 2 Design method of smiley

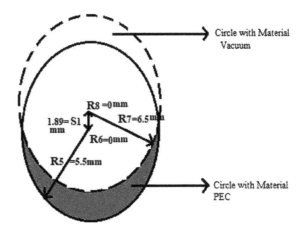

= 0.7 mm * 0.5 mm), dumbbell gap (D_g = 1.3 mm) and dumbbell bridge (D_b = 0.2 mm * 1 mm).

The structure of the smiley is shown in Fig. 2, starting with circular ring of PEC material. This had radii, outer radius R_5 = 5.5 mm, inner radius R_6 = 0. The other circular ring of vacuum area has radii outer R_7 = 6.5 mm and inner R_8 = 0.

Here, the dotted circle represents the vacuum, and the solid line represents the PEC materials. The two circles are separated with the distance S_1 = 1.89 mm. If the white portion of the PEC material is removed, then the required smiley would be created. This smiley is created on the upperpart of the substrate, and the difference between smiley and the second semicircle strip is 1 mm.

3 Designing Steps of Proposed Antenna

There are four steps to design the proposed antenna as follows:

(i) **Monopole antenna without DGS**
Here, Substrate (L, W, h) (mm) = (38, 52, 1.59), GND (L, W, h) (mm) = (11.5, 25, 0.5), Feed (L, W, h) (mm) = (12.06, 2.4, 0.5), Circle(I_r, O_r, h) (mm) = (8, 12, 0.5) and Semicircle (I_r, O_r, h) (mm) = (5.5, 9, 0.5).

(ii) **Monopole antenna with DGS**
Here, Substrate (L, W, h) (mm) = (38, 52, 1.59), GND (L, W, h) (mm) = (11.5, 25, 0.5), Feed (L, W, h) (mm) = (12.06, 2.4, 0.5), Circle(I_r, O_r, h) (mm) = (8, 12, 0.5), Semicircle (I_r, O_r, h) (mm) = (5.5, 9, 0.5), DGS: Dumbbell(L, W, h) (mm) = (0.7, 0.5, 0.5) and Bar(L, W, h) (mm) = (0.2, 1, 0.5).

(iii) **Monopole antenna with semicircular-shaped ring and DGS**
Here, Substrate (L, W, h) (mm) = (38, 52, 1.59), GND (L, W, h) (mm) = (11.5, 25, 0.5), Feed (L, W, h) (mm) = (12.06, 2.4, 0.5), Outer Semicircle (I_r, O_r, h) (mm) = (8, 12, 0.5), Inner Semicircle (I_r, O_r, h) (mm) = (5.5, 9, 0.5),

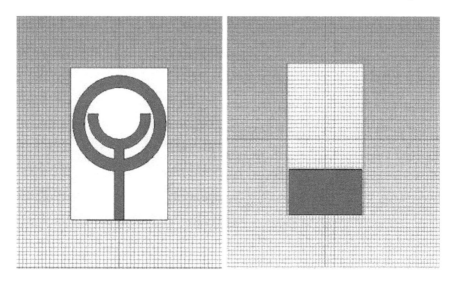

Fig. 3 Monopole antenna without DGS front and back view, respectively

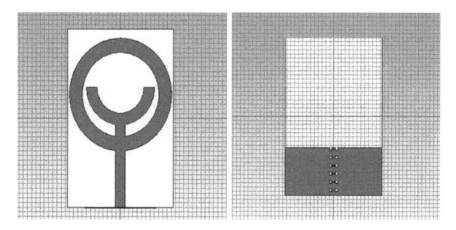

Fig. 4 Monopole antenna with DGS front and back view, respectively

DGS: Dumbbell(L, W, h) (mm) = (0.7, 0.5, 0.5), Bar(L, W, h) (mm) = (0.2, 1, 0.5) and Smiley (I_r, O_r, h) (mm) = (7.39, 8.28, 0.5).

(iv) **Monopole antenna with semicircular-shaped ring, smiley and DGS (proposed antenna)**

Here, Substrate (L, W, h) (mm) = (38, 52, 1.59), GND (L, W, h) (mm) = (11.5, 25, 0.5), Feed (L, W, h) (mm) = (12.06, 2.4, 0.5), Outer Semicircle (I_r, O_r, h) (mm) = (8, 12, 0.5), Inner Semicircle (I_r, O_r, h) (mm) = (5.5, 9, 0.5), DGS: Dumbbell (L, W, h) (mm) = (0.7, 0.5, 0.5), Bar(L, W, h) (mm) = (0.2, 1, 0.5) and Smiley (I_r, O_r, h) (mm) = (7.39, 8.28, 0.5).

Fig. 5 Monopole antenna with semicircular-shaped ring and DGS front and back view, respectively

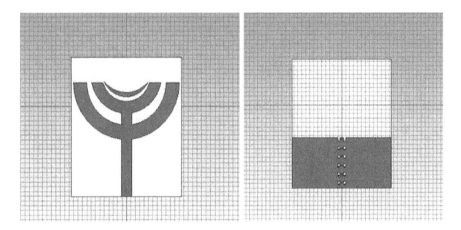

Fig. 6 Front and back view of proposed antenna

4 Results and Discussion

Impedance Bandwidth of Monopole Antenna without DGS:

The impedance bandwidth of monopole antenna without DGS using CST designed to resonate is shown in Fig. 7. The designed antenna resonates at 3.2 and 8.4 GHz, which covers the minimum required value of return loss of −10 dB.

Impedance Bandwidth of Monopole Antenna with DGS:

The impedance bandwidth of a cylindrical DRA using CST designed to resonate is shown in Fig. 8. The designed antenna resonates at 3.3 and 8.39 GHz, which covers the minimum required value of return loss of −10 dB.

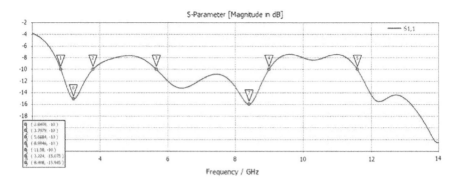

Fig. 7 Impedance bandwidth of monopole antenna without DGS

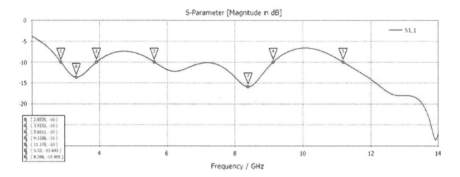

Fig. 8 Impedance bandwidth of monopole antenna with DGS

Impedance bandwidth of monopole antenna with semicircular-shaped ring and DGS:

The impedance bandwidth of monopole antenna with semicircular-shaped ring and DGS using CST designed to resonate is shown in Fig. 9. The designed antenna resonates at 6.02 GHz, which covers the minimum required value of return loss of −10 dB.

Impedance bandwidth of monopole antenna with semicircular-shaped ring, smiley and DGS:

The impedance bandwidth of monopole antenna with semicircular-shaped ring, smiley and DGS using CST designed to resonate is shown in Fig. 10. The designed antenna resonates at 6.06 GHz, covering the minimum required value of return loss of −10 dB.

(1) **Voltage Standing Wave Ratio**:

The ratio of the maximum voltage to the minimum voltage in a standing wave is known as voltage standing wave ratio. The term, which indicates the impedance

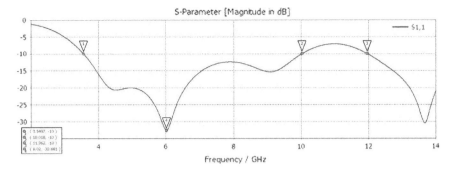

Fig. 9 Impedance bandwidth of monopole antenna with semicircular-shaped ring and DGS

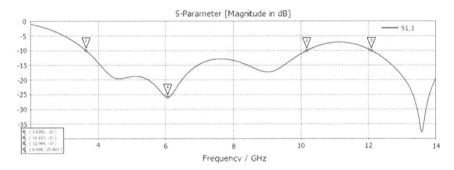

Fig. 10 Impedance bandwidth of monopole antenna with semicircular-shaped ring, smiley and DGS

mismatch, is VSWR. The higher the impedance mismatch, the higher will be the value of VSWR. The ideal value of VSWR should be 1 for effective radiation. The VSWR for different designs is shown below.

VSWR of monopole antenna without DGS

The VSWR of monopole antenna without DGS is shown in Fig. 11.

VSWR of monopole antenna with DGS

The VSWR of monopole antenna with DGS using CST is shown in Fig. 12.

VSWR of monopole antenna with semicircular-shaped ring and DGS

The VSWR of monopole antenna with semicircular-shaped ring and DGS is shown in Fig. 13.

VSWR of monopole antenna with semicircular-shaped ring, smiley and DGS

The VSWR of monopole antenna with semicircular-shaped ring, smiley and DGS is shown in Fig. 14.

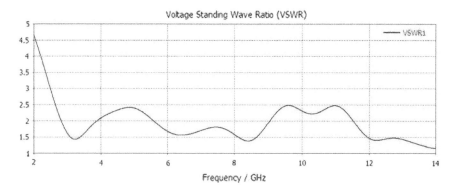

Fig. 11 VSWR of monopole antenna without DGS

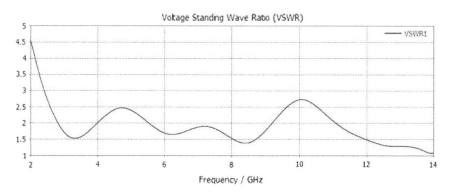

Fig. 12 VSWR of monopole antenna with DGS

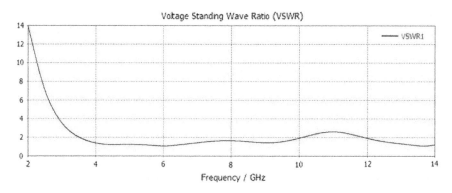

Fig. 13 VSWR of monopole antenna with semicircular-shaped ring and DGS

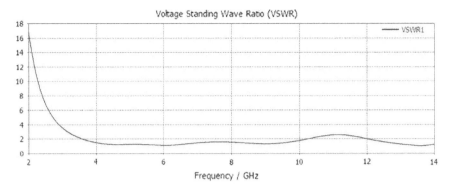

Fig. 14 VSWR of monopole antenna with semicircular-shaped ring, smiley and DGS

(1) **Directivity of an antenna**:

The ratio of maximum radiation intensity of the subject antenna to the radiation intensity of an isotropic or reference antenna, radiating the same total power is called the directivity.

Directivity of monopole antenna without DGS:

The radiation pattern showing the directivity (3D view) for the designed antenna and polar plots has been shown in Fig. 15 at 3.2 and 8.4 GHz, respectively.

Directivity of monopole antenna with DGS:

The radiation pattern showing the directivity (3D view) for the designed antenna and polar plots has been shown in Fig. 16 at 3.32 and 8.39 GHz.

Directivity of monopole antenna with semicircular-shaped ring and DGS:

The radiation pattern showing the directivity (3D view) for the designed antenna and polar plots has been shown in Fig. 17 at 6.02 GHz.

Directivity of monopole antenna with semicircular-shaped ring, smiley and DGS:

The radiation pattern showing the directivity (3D view) for the designed antenna and polar plots has been shown in Fig. 18 at 6.06 GHz.

Gain of an antenna:

Gain of an antenna is the ratio of the radiation intensity in a given direction to the radiation intensity that would be obtained if the power accepted by the antenna was radiated isotropic ally. The term antenna gain describes how much power is transmitted in the direction of peak radiation to that of an isotropic source. Gain is usually measured in dB. Unlike directivity, antenna gain takes the losses that occur also into account and hence focuses on the efficiency.

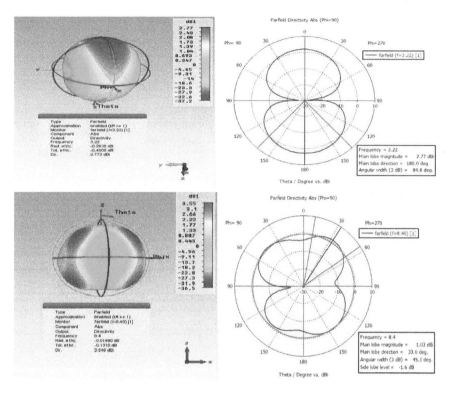

Fig. 15 Directivity of monopole antenna without DGS at 3.2 and 8.4 GHz (3D view and polar plot)

Gain of monopole antenna without DGS:

The gain in 3D and polar plots of monopole antenna without DGS is shown in Fig. 19.

Gain of monopole antenna with DGS:

The gain in 3D and polar plots of monopole antenna with DGS is shown in Fig. 20.

Gain of monopole antenna with semicircular-shaped ring and DGS:

Gain in 3D and polar plots of monopole antenna with semicircular-shaped ring and DGS is shown in Fig. 21.

Gain of monopole antenna with semicircular-shaped ring, smiley and DGS is shown in below Fig. 22.

5 Conclusion

A novel design of circular monopole antenna has been discussed in this work. The characteristics of simulation of the proposed antenna are matching with standards of

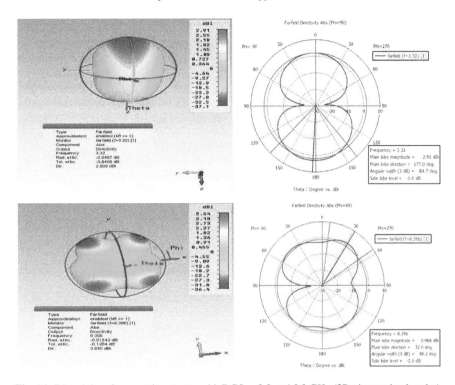

Fig. 16 Directivity of monopole antenna with DGS at 3.3 and 8.3 GHz (3D view and polar plot)

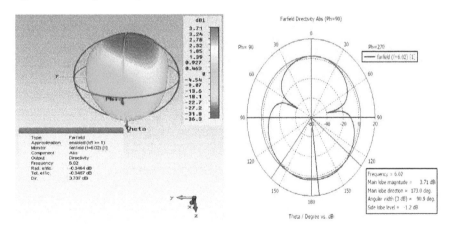

Fig. 17 Directivity of monopole antenna with semicircular-shaped ring 6.02 GHz (3D view and polar plot)

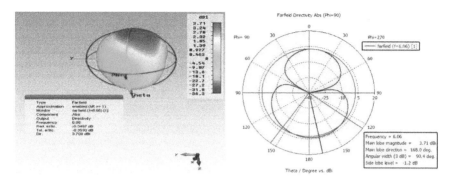

Fig. 18 Directivity of monopole antenna with semicircular-shaped ring, smiley and DGS at 6.06 GHz (3D view and polar plot)

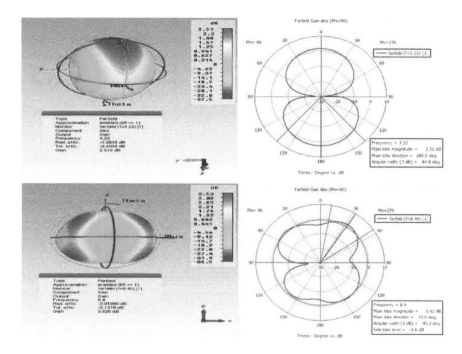

Fig. 19 Gain of monopole antenna without DGS at 3.2 GHz and 8.4 GHz (3D view and polar plot)

CC. This model is suitable for many short range usages. The omnidirectional characteristics in the operational frequencies and suitable matching of impedance render the structure appropriate for desirable band of utilization in the field of communication. The comparison of different design antennas is given in Table 2.

DGS-Based Wideband Microstrip Antenna for UWB Applications 193

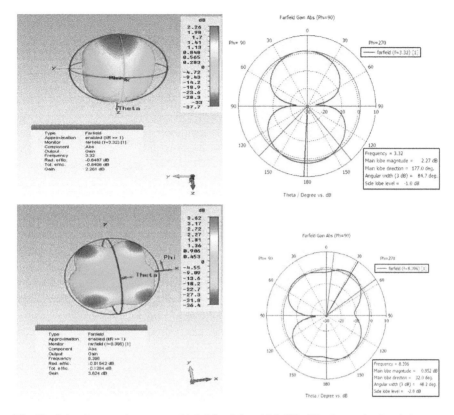

Fig. 20 Gain of monopole antenna with DGS at 3.3 and 8.3 GHz (3D view and polar plot)

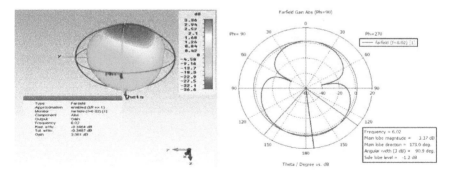

Fig. 21 Gain of monopole antenna with semicircular-shaped ring and DGS at 6.02 GHz (3D view and polar plot)

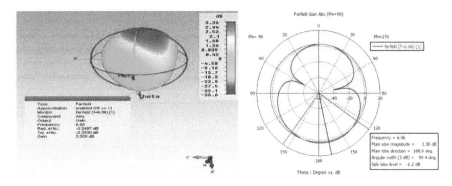

Fig. 22 Gain of monopole antenna with semicircular-shaped ring, smiley and DGS at 6.06 GHz (3D view and polar plot)

Table 2 Performance comparison of different antennas

Name of the antenna	Monopole antenna without DGS		Monopole antenna with DGS		Monopole antenna with semicircular-shaped ring and DGS	Monopole antenna with semicircular-shaped ring, smiley and DGS
Resonating frequency (GHz)	3.2	8.4		8.3	6.02	6.06
Bandwidth (GHz)	0.95	3.32	1.06	3.51	6.4	6.52
Gain (dB)	2.51	3.53	2.26	3.62	3.36	3.36
Directivity (dBi)	2.77	3.55	2.91	3.62	3.71	3.71

References

1. S.S. Mohan Reddy, P. Mallikarjuna Rao, B. Prudhvi Nadh, B.T.P. Madhav, K. Aruna Kumari, Design and Analysis of Compact Circular Half-Ring Monopole Antenna with DGS for UWB Applications
2. C.P. Lee, C.K. Chakrabarty, Ultra wideband microstrip diamond slotted patch antenna with enhanced bandwidth. Int. J. Commun. Netw. Syst. Sci. vol. **4**(7), 468–472 (2011)
3. K.S. Ritu, Microstrip antenna design for UWB applications. Int. J. Adv. Res. Comput. Commun. Eng. **2**(10), 3824–3828 (2013)
4. Y. Singh, D. Singh, G.S. Gill, Design of wideband microstrip antenna for UWB applications. Int. Res. J. Eng. Technol. **2**(5), 995–998 (2015)
5. B.T.P. Madhav, H. Kaza, Novel printed monopole trapezoidal notch antenna with S-band rejection. J. Theor. Appl. Inf. Technol. **76**(1), 42–49 (2015)
6. P. Singh, R. Tomar, The use of defected ground structures in designing microstrip filters with enhanced performance characteristics. Proc. Technol. **17**, 58–64 (2014)
7. S.S. Mohan Reddy, P. Mallikarjunarao, B.T.P. Madhav, Asymmetric defected ground structured monopole antenna for wideband communication systems. Int. J. Commun. Antenna Propag. **5**(5), 256–262 (2015)

8. L.H. Weng, Y.C. Guo, X.W. Shi, X.Q. Chen, An overview on defected ground structure. Progr. Eletromagn. Res. **B7**, 173–189, (2008)
9. J. Pei et al., Miniaturized triple-band antenna with a defected ground plane for WLAN/WiMAX applications. IEEE Antennas Wirel. Propag. Lett. **10**, 298–301 (2011)
10. N. Ahuja, R. Khanna, J. Kaur (2012) Dual band defected ground microstrip patch antenna for WLAN/WiMax and satellite application. Int. J. Comput. Appl. **48**(22), 1–5 (2012)
11. FCC First report and order on ultra wide band technology, Washington, DC (2002)
12. P.S. Sundar, S.K. Kotamraju, T.V. Ramakrishna, B.T.P. Madhav, Novel miniatured wide band annular slot monopole antenna. Far East J. Electron. Commun. vol. **14**(2), 149–159 (2015)
13. K.V.L. Bhavani, H. Khan, B.T.P. Madhav (2015) Multiband slotted aperture antenna with defected ground structure for C and X-band communication applications. J. Theor. Appl. Inf. Technol **82**(3), 454–461 (2015)

Brain Tumor Segmentation Using Fuzzy *C*-Means and Tumor Grade Classification Using SVM

V Ramakrishna Sajja and Hemantha Kumar Kalluri

1 Introduction

The brain could be a complicated and vital organ, and treatment usually causes life-long changes. A brain tumor happens once there is an irregular enlargement of cells called tumor, within the neural structure. Essential brain tumors might be considerate or harmful. A benign tumor increment bit by bit has particular points of confinement and on occasion spreads. A malignant cerebrum tumor overgrows, has unusual breaking points, and spreads to neighboring regions. Disregarding the way that they are sometimes called cerebrum development, desperate mind tumors do fit the significance of harm, since they do not spread to organs outside the mind.

The World Health Organization (WHO) developed a classification and grading system for regular communication, treatment planning and analyzed outcomes for MR brain tumors. Magnetic resonance imaging [1] is employed to look at the brain life structure. It is more practical than computed tomography (CT). Tumors are classified by their cell kind and grade by viewing the cells. The category describes the method tumor cells look beneath the magnifier and is a sign of aggressiveness (low grade suggests that least aggressiveness and high grade suggests that most aggressiveness). Tumors usually have a mixture of cell grades and may change as they grow.

Grade 1 and 2 tumors [2] are low grades, slow-growing, comparatively contained and unlikely to spread to different components of the brain. There is additionally less probability of them returning if they will be fully removed. They are typically still mentioned as benign. The term benign is less used these days as this may be misleading. These low-grade brain tumors will still be serious. This is because of

---

V. R. Sajja (✉) · H. K. Kalluri
Department of CSE, VFSTR Deemed to Be University, Vadlamudi, AP, India
e-mail: vramakrishnasajja@gmail.com

H. K. Kalluri
e-mail: hemanth_mtech2003@yahoo.com

© Springer Nature Singapore Pte Ltd. 2020
J. Fiaidhi et al. (eds.), *Smart Technologies in Data Science and Communication*, Lecture Notes in Networks and Systems 105, https://doi.org/10.1007/978-981-15-2407-3_24

the development will cause harm by going ahead and harming close regions of the mind, because of the restricted zone limit of the skull. They willl moreover block the stream of the cerebrospinal fluid (CSF) that sustains and protects the brain, causing the growth of pressure on the mind.

Grade 3 and 4 tumors [3] are high grade, quick growing and maybe said as malignant or cancerous growths. They will probably spread to various segments of the cerebrum and may return, regardless of whether seriously treated. They cannot sometimes be treated by surgery alone, however usually needs different treatments, such as radiotherapy and chemotherapy.

In this paper, a computerized methodology presented which classifies the tumor whether it is normal tumor or abnormal tumor. The proposed methodology includes several stages like preprocessing, segmentation, feature extraction, and classification. Several data mining techniques [4] like fuzzy C-means algorithm and support vector machine are used at respective stages to develop the proposed methodology.

The rest of the paper is categorized as follows. Section 2 describes the related work, Sect. 3 describes the proposed methodology; Sect. 4 is about data set and experimentation results. Finally, Sect. 5 is about conclusion and future scope.

2 Related Work

The support vector machine is a famous classification algorithm, and it was used by many researchers. Nandpuru et al. [5] have done work on support vector machine for detection of brain tumors using MRI images. In this paper, they extracted the features of a grayscale image, and those were given to SVM to classify, which could produce the accuracy of 74%.

Padma et al. [6] have done a work support vector machine for detection of brain tumors using wavelet features. They have used CT images as input images and stressed the efficiency of CT images in the field of pathology. The researchers reported accuracy of 93.3%.

Gupta et al. [7] have applied preprocess techniques to enhance the image, fuzzy C-means algorithm was employed to isolate the tumor region, and then gray-level run length matrix was used to extract the features. GLRLM features are given as input to SVM classifier to classify the tumor which could produce the accuracy of 91.77%.

3 Proposed Work

A computerized methodology is presented in Fig. 1, which classifies the tumor type whether it is normal tumor or abnormal tumor. The proposed method includes several stages like preprocessing, segmentation, feature extraction, and classification.

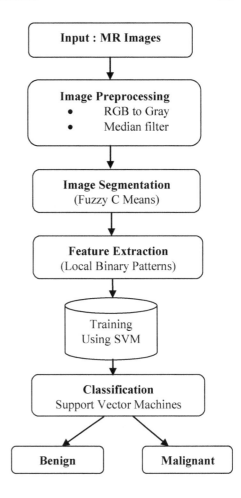

Fig. 1 Block diagram of the proposed architecture

Several data mining techniques like fuzzy C-means algorithm and support vector machine are used at respective stages to develop the proposed methodology.

A. **Preprocessing**

Preprocessing is an essential step to enhance the quality of image going to be considered for further examination. If preprocessing is absent, then results are not anticipated as good. In this work, collected images are in RGB color, and those are converted from color image to grayscale image. Grayscale image is given as input to the median filter to get the output as smoothened image. This enhanced image [8] will be given as input to the segmentation phase.

B. **Segmentation using Fuzzy C-Means**

Segmentation [9] is the second phase of the proposed methodology. The fuzzy C-means algorithm is employed to segment the tumor region which could provide

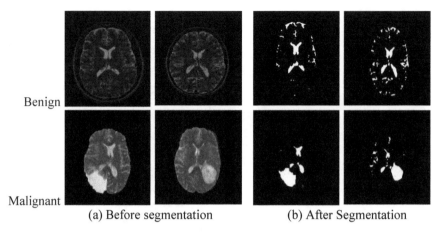

Benign

Malignant

(a) Before segmentation (b) After Segmentation

Fig. 2 Results of fuzzy C-means segmentation. **a** before segmentation, **b** after segmentation

intelligent artifacts. Here, data is segmented into portions and experimentation results of fuzzy C-means algorithm are given in Fig. 2.

Fuzzy C-means segmentation algorithm:

Step 1. Consider the data points (for example, four points) in two-dimensional space as column-wise data.
Step 2. Initialize the number of clusters 'c' and set criteria for convergence of $\epsilon_L = 0.01$.
(i.e.,) $\max_{ik} \left| \mu_{ik}^{(r+1)} - \mu_{ik}^{(r)} \right| \leq 0.01$
Step 3. Initialize membership metrics with an assumption of data points 1–3 are in cluster 1, and fourth data point is in cluster 2.

$$\mu^0 = \begin{bmatrix} 1 & 1 & 1 & 0 \\ 0 & 0 & 0 & 1 \end{bmatrix} \begin{matrix} C_1 \\ C_2 \end{matrix}$$

Step 4. Calculate the initial cluster center using Eq. 1

$$V_{ij} = \frac{\sum_{k=1}^{n} (\mu_{ik})^2 X_{kj}}{\sum_{k=1}^{n} (\mu_{ik})^2} \quad \text{for} \quad i = j = 1 \text{ to } c \qquad (1)$$

Step 5. Calculate the distance d_{ik} (for $i = 1$ to 'c' and $k = 1$ to 'n') of each data point from each cluster.
Step 6. Update membership metrics using Eq. 2

$$\mu_{ik}^{(r+1)} = \left[\sum_{j=1}^{c} \left(\frac{d_{ik}^{(r)}}{d_{jk}^{(r)}} \right) \right]^{-1} \quad (2)$$

Step 7   Check for convergence (maximum absolute value of pairwise comparison of each value in $\mu^{(r)}$ and $\mu^{(r+1)}$) using Eq. 3

$$\max_{ik} \left| \mu_{ik}^{(r+1)} - \mu_{ik}^{(r)} \right| \leq 0.01 \quad (3)$$

Step 8. Repeat steps 4–7 until no change in cluster centers or satisfying the convergence.

C. Feature extraction using LBP

Local Binary Pattern (LBP) is a traditional algorithm to extract the features which could be useful to classify the tumor type. In this study, 255 features are extracted for each image of 288 images includes normal and abnormal categories. Resultant features are given as input to the classification phase to classify the tumor grade.

D. Classification Using SVM

The SVM approach depends on the analysis of a supervised learning algorithm and is employed to one-class categorization issues to n-class categorization issues. The foremost goal of the SVM [10] approach is to convert a nonlinear separating element to a linear conversion by means of alteration using a function known as SVM's kernel function. Support vector machine will yield good results by varying the decision planes. A decision plane partition the set of data points into two classes based on their class memberships. The classification phase will be finished successfully only upon the identification of the correct tumor class. The implementation of SVM includes two essential stages of testing and training. In support vector machine, the classes have expected to be recognized as ± 1, and the resultant border is projected as $y = 0$. w is the weight vector, b the offset. Since the classes were well-defined as ± 1, the equation for the line separating the classes is specified in Eqs. 4 and 5.

$$x_i w + b \geq 1 \quad \text{when} \quad y = +1 \quad (4)$$

$$x_i w + b \leq 1 \quad \text{when} \quad y = -1 \quad (5)$$

The distance from the hyperplane to the origin is $M = \frac{2}{\|w\|}$, where M is the margin. The distance from the hyper-plan ($x_i w + b = 0$) to the origin is $\frac{-b}{\|w\|}$, where $\|w\|$ is the norm of w, so the maximum margin is obtained by minimizing $\|w\|$.

4 Dataset Preparation and Experimentation Results

Several open databases provide the statistics required for this research. The proposed approach has been validated on BRATS 2018 dataset. BRATS 2018 dataset consists of 108 normal images and 180 abnormal images. To train the SVM classifier, 230 images are used including 87 normal images and 143 abnormal images. Similarly, to test the classifier, 58 images are used including 21 normal images and 37 abnormal images. Dataset details are depicted in Table 1.

To test the performance of the SVM classifier, authors have constructed confusion matrix. Confusion matrix has four important terms labeled as TP, TN, FP, and FN and those can be used in computing many evaluation measures. Those four terms are also called as building blocks for any classifier. These terms are summarized in the confusion matrix of Table 2.

TP is pronounced as true positive which refers to the positive tuples those are rightly classified. TN is pronounced as true negative which refers to the negative tuples those are correctly classified. FP is pronounced as false positive which refers to the negative tuples those are wrongly classified. FN is pronounced as false negative which refers to the positive tuples those are wrongly classified.

Accuracy of the classifier is computed using metric accuracy which calculates the ratio of the total number of correct predictions from whole data by using Eq. 6.

$$\text{Accuracy} = \frac{\text{TN} + \text{TP}}{\text{FP} + \text{FN} + \text{TP} + \text{TN}} \qquad (6)$$

The error rate is also called a misclassification rate of a classifier. It is computed as simply 1-accuracy, and it is determined by using Eq. 7.

$$\text{Error Rate} = \frac{\text{FP} + \text{FN}}{\text{TP} + \text{TN} + \text{FP} + \text{FN}} \qquad (7)$$

Table 1 The MRI brain dataset used for the proposed model

BRATS 2018	Normal images	Abnormal images	Total
Training	87	143	230
Testing	21	37	58
Total	108	180	288

Table 2 Confusion matrix

Class	Yes	No	Total
Yes	TP	FN	P
No	FP	TN	N
Total	PI	NI	P + N

True positive rate is also called as sensitivity or recall. It computes the ratio of true positive tuples from total positive tuples and it is determined by using Eq. 8.

$$\text{True Positive Rate} = \frac{TP}{TP + FN} \qquad (8)$$

True negative rate is also called as specificity. It computes the ratio of true negative tuples from total negative tuples, and it is determined by using Eq. 9.

$$\text{True Negative Rate} = \frac{TN}{TN + FP} \qquad (9)$$

After performing the experimentation on the considered dataset according to the proposed architecture, it produces 94.8% accuracy, 5.2% error rate, 90.90% recall, and 97.2% specificity. Resultant values for building blocks of the classifier such as TP, FP, FN, and TN are shown in Table 3.

The confusion matrix describes the results of the test data. Initially, SVM classifier was trained with 230 images than after 58 images of normal and abnormal were tested by using same classifier. Classifier produces the output as 20 images in true positive category, 35 images in true-negative category, 1 image in false-positive category and 2 images in the false-negative category. The various evaluation measures of proposed classifier and existing classifier are compared, and those results are given in Table 4.

The proposed method exhibits more accuracy, recall, and specificity over the existing classifier. Due to the integrated architecture of proposed method, it performs extraordinary results over existing classifiers. Figure 3 depicts the performance of existing classifier and proposed classifier. The accuracy of the proposed classifier is 94.8% when it is compared to existing classifier which holds the accuracy of 91.77%.

Table 3 Confusion matrix for the proposed architecture

Class	Yes	No	Total
Yes	20	2	22
No	1	35	36
Total	21	37	58

Table 4 Comparative study on evaluation measures

Evaluation measures	Proposed classifier (SVM with LBP features) (%)	Existing classifier (SVM with GLRLM features) (%)
Accuracy	94.8	91.77
Error rate	5.2	8.23
True positive rate, sensitivity, recall	90.90	88.45
True negative rate, specificity	97.2	100

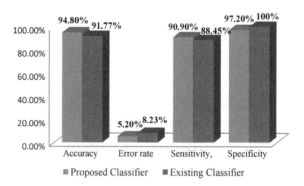

Fig. 3 Comparative analysis of existing and proposed classifiers

5 Conclusion and Future Scope

This paper proposed method has been validated on the benchmark data sets and overwhelms that it produces good results in detection of brain tumors. The integrated architecture includes fuzzy C-means clustering and support vector machine classification returns accurate performance in brain tumor identification. In future work, performance rate can be increased, and error rate can be decreased by trailing different segmentation algorithms and also by varying the kernels in SVM classifier.

References

1. R.H. Caverly, MRI fundamentals: RF aspects of magnetic resonance imaging (MRI). IEEE Microw. Mag. **16**(6) (2015)
2. V. Ramakrishna Sajja, G.R. Nitta, Experimental approach for detection of brain tumor grade using SVM classification. Int. J. Recent Technol. Eng. (IJRTE) **7**(5S4). ISSN: 2277-3878 (2019)
3. V. Ramakrishna Sajja, S.R. Rani, D.S. Bhupal Naik, K. Pratyusha, Segmentation of Brain tumor using hybrid approach of fast bounding box and thresholding in MRI. Int. J. Recent Technol. Eng. (IJRTE) **7**(5S4). ISSN: 2277–3878 (February 2019)
4. M. Hanumanthappa et al., Data mining in healthcare: a survey of techniques and algorithms with its limitations and challenges. IJERA **3**(6), 937–941. ISSN: 2248–9622, Nov–Dec 2013
5. H.B. Nandpuru et al., Magnetic resonance imaging brain cancer classification using SVM, in *2014 IEEE Students 'EECS Conference*. 978-1-4799-2526-1/14/$31.00 ©2014 IEEE (2014)
6. R. Sukanesh et al., Automatic classification and segmentation of brain tumor in CT images using optimal dominant gray level run length texture features. MJSR **13**(7), 883–888 (2013)
7. G. Gupt et al., Brain tumor segmentation and classification using FCM and support vector machine. IRJET **4**(5), 792–796. e-ISSN: 2395–0056, May 2017
8. N.M. Kundeti, H.K. Kalluri, S.V. Rama Krishna, Image enhancement using DT-CWT based cycle spinning methodology, in *IEEE International Conference on Computational Intelligence and Computing Research*, 978-1-4799-1597-2 (2013)
9. V. Ramakrishna Sajja, N. Gnaneswara Rao, Segmentation of medical images using enhanced K-means, wavelet transformations, and morphological operations, in *CIE46 Proceedings*, 29–31 Oct 2016, Tianjin/China (2016)
10. B. Gupta, et al., Brain tumor detection using support vector machine and curvelet transform. IJCM **3**(4), 1259–1264. ISSN 2320–088X (2014)

Optimized Water Scheduling Using IoT Sensor Data in Smart Farming

Kolli Venkatra Krishna Kishore, B. Yaswanth Kumar
and S. Venkatramaphanikumar

1 Introduction

Water preservation has been of esteem importance worldwide. It is not merely for just the environmental factors but for the existence of life on earth. Water is the source of life as many scientific studies conclude, water has been deem recognized in the evolution of human civilization and many other species as well. Thus, the other most important living beings that require water besides being the main source of food are plants; water is one of the necessary factors for growth of plants. Plants require water as essentially as sunlight for performing photosynthesis. Not many plants sustain with high saline levels in ocean water, and hence, fresh water is needed for irrigation purposes. Precision agriculture forms the basis for irrigation water management that aims to preserve water by regulating the frequency of watering the plants. As 70% of all water is consumed in agricultural processes, by carefully calibrating the agricultural practices with the need of the crop, preservation of water for future generations is not a dream that cannot be realized.

The heart of precision agriculture lies in the sensor nodes used. The values obtained from various sensors are taken into consideration to make informed decisions, to get a high yield with optimum use of water. One such parameter is soil moisture content. It is defined as the water available in the top 10 cm of the soil, whereas the water content from top 200 cm is categorized as root zone soil moisture. Many of the evolving as well as pre-existing techniques use soil moisture rather than root zone soil moisture as a parameter, as it becomes cumbersome to place the sensors by digging the soil for every plant.

K. Venkatra Krishna Kishore · S. Venkatramaphanikumar (✉)
Department of Computer Science and Engineering, Vignan's Foundation for Science, Technology and Research, Vadlamudi, Guntur, Andhra Pradesh, India
e-mail: svrphanikumar@yahoo.com

B. Yaswanth Kumar
Department of Information Technology, Vignan's Foundation for Science, Technology and Research, Vadlamudi, Guntur, Andhra Pradesh, India

With recent technologies at hand, remote sensing of parameters had become a task of no time. Besides, it is more efficient and there is no need for human interaction, while traditional methods use stress water index values derived from neutron probes or tensiometers or gypsum blocks.

With enough said, the proposed work is aimed at reducing water wastage in the major sector of water usage, i.e., agriculture systems. By obtaining physical parameters as soil moisture, soil temperature, air temperature, air humidity and UV index from the sensors, the sensor data are to be analyzed by a support vector regression algorithm to bring out the relations between them and help the system to have the capability to take informed decisions whether to start or stop irrigation cycle and also to plan the irrigation cycles whenever required.

Section 2 is about the literature study required for forming this complex work; this chapter gives details about the pre-existing work done in individual segments that all together form a new proposed system. Section 3 deals with the introduction of the proposed architecture and defines a brief overview of the usage of module a specified purpose in the proposed work. Implementation of the proposed system is presented in Sect. 4, which gives the detailed reasoning of usage of each component and gives the explanation of how the work has been streamlined to arrive at a successful real-time implementation of the proposed work. The results and conclusion part are presented in Sect. 5 to understand the clear outcomes of the project and also gives insights into what future work inspires to be done yet to achieve a near-perfect system.

2 Related Work

Harris Drucker et al. [4] proposed that support vector regression (SVR) has the highest capability of handling problems with high dimensionality owing to that optimization does not depend on input dimensionality, whereas feature space selection has high capability of handling less dimensionality. SVR was compared to bagging and feature space representation on different nonlinear problems; feature representation came to be the best, SVR came at the second, while bagging stood at the last place. Boston Housing used Support Vector Regression to realize optimum results, while feature space representation was unable to construct high dimensionality. Hence, it was concluded that SVR can handle high-dimensionality input data with an ease. Gill et al. [5] have found SVR to be more efficient at predicting soil moisture when compared to artificial neural networks (ANNs). In this work, soil moisture and meteorological data were considered for $t - 1$ and t days, respectively, and the soil moisture was predicted for $t + 4$ and $t + 7$ days, respectively. It was observed that the combination of previous knowledge of soil moisture data and current meteorological data finds the best correlation between the estimated values and actual measurements. While consideration of one of the parameters has produced not much better results, with only meteorological data used, slight degraded correlation is observed, while inclusion of soil moisture data has turned out to be the least accurate of the three combinations. It has been observed that SVR performs better in all cases when compared with

ANN. Davis et al. [3] had proposed that evapotranpiration (ET) controllers were found to save water substantially even when compared to time clock scheduled water saving methods. The study comprised a comparative differentiation of water saving capabilities of three ET controllers and time clock-based TIME and RTIME controllers used in household. Time-based scheduling water saving was taken into consideration without a rain sensor. Nonetheless, the study period was drier than 30-year historical average rainfall. ET controllers have recorded 43% of average water savings compared to time-based treatment without rain sensor; it was almost twice effective when compared with a rain sensor. Maximum savings of 60% were measured during winter 2006–07, and minimum savings of 9% were recorded during spring 2007. Time-based treatment had shown similar water savings in fact that regular adjustment of clocks is required to the individuals. Hence, ET controllers would offer promising water savings if programmed correctly.

Goldstein et al. [6] proposed that tree-based machine learning algorithms performed better at predicting irrigation patterns. Gradient boosting regression (GBR) tree was able to predict irrigation patterns, by utilizing data from parameters such as soil, weather and historical yield for the crop, with an accuracy of 93%, while boosted tree classifier was able to predict patterns with an accuracy of 95%. Linear regression was the least performer of the three. The proposed work aims at automation of the work done by an agronomist, determining the weekly irrigation plan with the use of sensor data from the plant and weather and irrigation data. However, attempts are not made to extract the relationship between the data or to determine whether there exist some interesting patterns. Hence, the proposed work was developed with data from the past two years and different regression and classification algorithms were applied to predict a weekly irrigation plan as an agronomist would do. Simply put, regression and classification algorithms are used to replicate the manual work done by an agronomist, to determine a water scheduling plan for the plants. While GBR had slight accuracy degradation over BTC, GBRT is most suitable for agriculture pattern predictions since BTC can only classify data into classes.

Viani et al. [10] presented a decision support system with the combination of wireless sensor actuation network and fuzzy logics, with the farmer inputs from water scheduling; this is optimized in irrigation and reduces the water wastage. The model was intended to balance water content and simulate the reaction of soil to the proposed system. Parameters such as water precipitation, evapotranspiration [1] and percolation have been considered the inputs with threshold-based and multiple-threshold-based approach; the threshold-based approach proved to be more conservative with water. Fuzzy logic calibration has been done with inputs from the farmer to duplicate the human experience and to understand the state of the crop. Integration of WSAN architecture with FL-based decision support system has advantageous features such as improved water saving and reduced water level stress index.

3 Proposed Architecture

The proposed architecture consists of the modules such as sensor data pre-processing, soil moisture prediction and motor pumping, and all the modules are well connected via Internet access. The sensor data are obtained from the field sensors via the object node, i.e., the Arduino microcontroller. The collected data are sent to the controller node of Raspberry Pi via Internet protocols [7] and [8]. Further, the received data were sent to controller node to the soil moisture predicting algorithm situated in the cloud. The cloud then receives the data, stores them in the database, performs regression and clustering algorithms and predicts the soil moisture for upcoming days. The value thus obtained is then related to the current soil moisture data, and then decision-making takes place in the cloud. The decision hence determines whether to start/stop irrigation. Finally, the decision is forwarded to the controller node, to start/stop irrigation by turning on/off the motor, which is connected to a relay switch with Wi-Fi-enabled Arduino (Fig. 1).

3.1 In-Depth View of the Proposed System

The in-depth view of the proposed architecture is presented in Fig. 2. In this, as mentioned earlier, both the sensor data module and cloud analysis module are well connected via Internet [11]. The sensor data can be easily understood in the bottom-up approach. At the bottom tier, field sensor and weather data modules are being placed and connected to the Internet. Weather data are transferred to the controller node via HTML commands, while field sensor data are obtained from the sensors connected to the object node, i.e., Arduino is set to the controller node via programming of the Arduino IDE and the HTML commands to the controller node. The controller node sends the received data to the cloud Web application setup via web services. The data received are stored in the database of the cloud system, and then, data are sent to the support vector regression algorithm for predicting soil moisture for 'n' days. The derived output is then forwarded for classifier to compute the prediction accuracy.

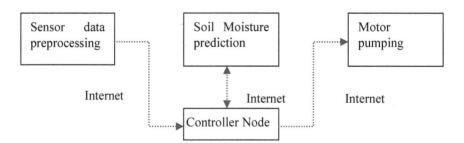

Fig. 1 Proposed system architecture

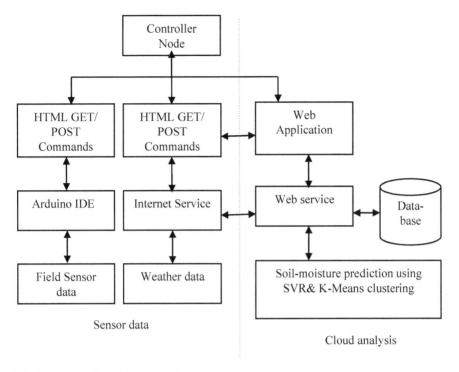

Fig. 2 In-depth view of the proposed system

Prediction of soil moisture in the proposed system makes the decision to set the irrigation date or the system checks for any rain forecast on irrigation date or nearby irrigation date, and then the system halts irrigation on the decided date and waits for rain to come. Besides, system continuously checks for the soil moisture content; if moisture levels fall below the threshold and are not able to maintain minimum soil moisture, then the proposed system starts the water pumping enough to maintain minimum moisture levels until rain arrives.

3.2 Support Vector Machines

In general, support vector machine (SVM) is a discriminating classifier, which defines an optimal hyperplane as shown in Fig. 3. These hyperplanes divide the data objects into two classes by defining a clear isolation. In the real scenario, data are prone to be present inside the other targeted data. A simple two-dimensional X-Y plane is not enough to resolve the scenario in Fig. 3b. Hence, consider an extra plane Z to map the data points using $Z = X2 + Y2$. The transformations described in Fig. 3a, b are not enough to find the circular boundaries described in Fig. 3c. Those transformations are defined as kernels. A similar yet more complex problem can arise when the data

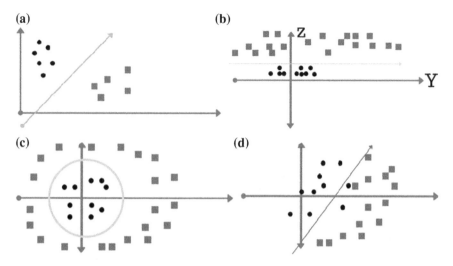

Fig. 3 Various kernels of support vector machine

are overlapped. Firstly, SVM follows the rule-based classification to maximize the margin among two classes as accurate as possible. Support vector regression is used to solve regression problem described in the proposed optimum water scheduling.

4 Proposed IoT-Based Optimized Water Scheduling

The proposed work serves in the combination of Arduino as an object node and Raspberry Pi as a controller node. The controller node sends the field sensor values to the cloud database to be stored there for further prediction of soil moisture values. Various field sensor data such as soil moisture, soil temperature, air humidity, air temperature and UV radiation are collected from respective sensors. The detailed methodology of the proposed work is presented in Fig. 4.

4.1 Arduino Uno

Arduino Uno is a microcontroller board, based on the ATmega328. It supports interfacing with various sensors to collect the data from sensors. Arduino can be connected to a computer to write the desired code for the microcontroller to work with. Arduino Uno consists of 14 digital output ports with 6 being capable of producing power width modulation, and it contains 6 analog input pins, ground pins to be connected to ground, with additional 3.3- and 5-V output pins. The ATmega328, an 8-bit

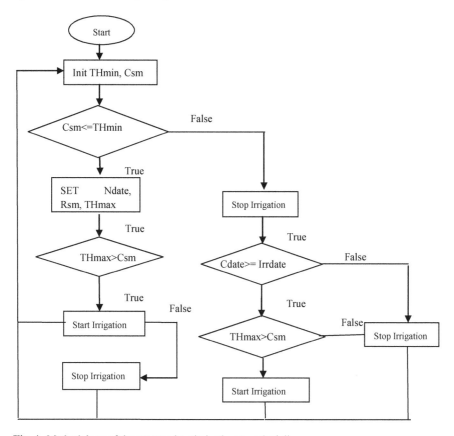

Fig. 4 Methodology of the proposed optimized water scheduling

microcontroller, features a useful power-save mode so that the power is not wasted when sensors are delayed for a longer time.

The power-save mode is useful when the board must be powered up with a battery, when deployed as a remote node (Fig. 5).

4.2 Details of Sensors Used in the Proposed Optimal Water Scheduling

The proposed system has been implemented using field sensors such as soil moisture, soil temperature, air temperature, humidity and UV intensity (Fig. 6).

DHT11 DHT11, a temperature and humidity sensor, is used to capture digital output values. It can measure values ranging from 20 to 90% of relative humidity and

Fig. 5 Arduino Uno microcontroller

Fig. 6 DHT11 sensor

temperature from 0 to 50 °C. It detects water vapor by the electrical resistance between two electrodes.

Soil Moisture Sensor The Soil moisture sensor is used to measure the water content (moisture) of soil. When the soil is having water below the threshold level determined by the user, the module output is at high level; else the output is at low level. The sensor has two electrodes which conduct electricity, and the resistance between the electrodes is inversely proportional to the moisture content of the soil. It translates to that low resistance derives high moisture content, while high resistance defines lower moisture levels in the soil (Fig. 7).

DS18B20 Soil Temperature Sensor DS18B20 is a waterproof temperature sensor, useful for measuring underwater temperatures. This sensor is used to measure the temperatures in the soil. The DS18B20 is a one-wire communication-based sensor, i.e., it requires only the data pin to be connected to the Arduino in series with a pull-up resistor, while the other pins are used to connect to power the module. This is a digital output sensor; the sensor can be connected to a microcontroller with the

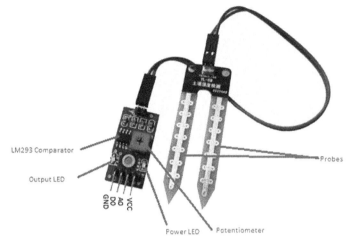

Fig. 7 Soil moisture sensor

data and ground pins connected, while there is no need for connecting to a input power supply (Fig. 8).

ML8511 UV Sensor The ML8511 UV sensor works by giving out an analog signal, in relation to the amount of ultraviolet light detected. Ultraviolet light radiation occurs from 10 to 400 nm wavelength in electromagnetic spectrum up above the earth's atmosphere. Warn the user to detect the UV index as it relates to weather conditions. The sensor measures the light most effectively in the range of 280–390 nm, categorized as part of the UVB (burning rays) spectrum and most of UVA (tanning rays) spectrum. The sensor outputs an analog voltage that is linear to the measured UV intensity (mW/cm^2). ML8511 comes with an internal amplifier; it converts photo-current sensed to voltage depending on the UV intensity (Fig. 9).

In the deployed system, a standalone object node is used as a controller node. For large farming area, a ZigBee scenario can be implemented with a controller node to

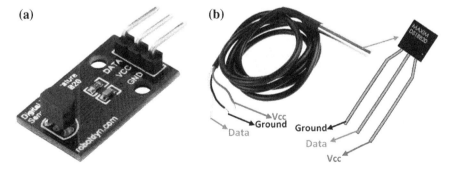

Fig. 8 **a** DS18B20 sensor with module, **b** module

Fig. 9 ML8511 UV sensor

communicate with multiple object nodes placed at different places in the field. These sensors are used along with an Arduino Uno, to obtain the sensor values periodically. The sensor values obtained are given to the support vector regression algorithm, and then the algorithm gives out the predicted soil moisture values of day 1, 2, ..., n. The obtained output, to obtain more accurate values, is further given as input to classification algorithm. The resultant output will be the accuracy of predicted soil moisture for next n days. The proposed work is implemented with a submersible 12-V pump motor for experimental convenience. In real-time scenario, a 1-khp motor can be made to turn on with a higher-capacity relay module, thus enabling groundwater to be pumped for irrigating the farm area.

4.3 Component Diagram of the Proposed Smart Water Scheduling

The effect of air temperature on the growth of the plant is correlational since an increase in temperature of air leads to an increase in temperature of soil, making the rate of transpiration to go high. By avoiding the effects of high air temperatures, plant growth can be enhanced accordingly for a good yield. The proposed architecture must be calibrated accordingly for the necessity of the plant, such that the unique rate of water consumption of the plant is addressed. The areas for irrigation scheduling are divided into sectors, with one sensing node sensing the values within that location. The values are further sent to the controller node to be sent to the cloud the database. Here, state is classified with the classifier model on the sensing data and is stored in database. The identified status is forwarded to the scheduler. Scheduler uses different queues for different classes and takes decision to release outlet of water based on the highest priority given to most dry sectors. Ideally, the most dry scheduler should always be empty. However, due to any external factors, if any sector is identified with most dry class, the scheduler must be given the highest priority to start irrigation for the sectors in the queue. The dry classified sectors are programmed to check for

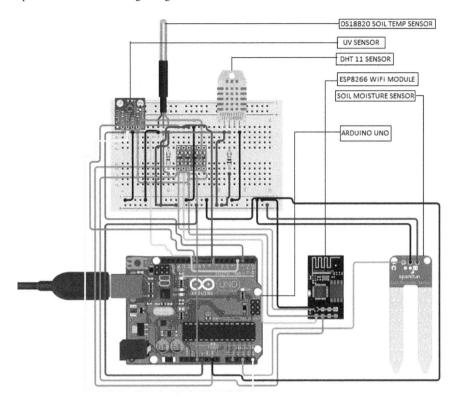

Fig. 10 Component diagram of the proposed methodology

irrigation scheduling by considering other parameters such as air temperature, soil temperature and air humidity values from the field sensor values (Fig. 10).

Soil moisture prediction algorithm:

1. Initialize the field sensors and start acquisition of data.
2. Train the model for prediction of soil moisture with the field sensor data.
3. Predict the variations in soil temperature using weather data.
4. Train the model for soil moisture prediction with weather data and output of previous model.
5. Predicted soil moisture difference (PSMD) for each next day $\{p_0, p_1, p_2, ..., P_n\}$ for day $\{1, 2, 3, ..., n\}$ is obtained.
6. For n number of days cluster using K-means on the output PSMD K-means on (SMD of S_d, P_i, NoC).
7. Finally, predicted values of soil moisture are obtained for each next day as $NPSMD_i$.
8. The soil moisture predicted will be used to water the plants accordingly.

Controller Algorithm

Initialize minimum threshold (Th_{min}) and maximum threshold (Th_{max}) of soil moisture to start and stop irrigation.

```
    Read the current soil moisture ( Csm)
    If(Csm< = THmin)
    {
Ndate = nearest date of precipitation from upcoming days.
    Required soil moisture (Rsm) = sum(predicted SMD from current date to nearest
    precipitation data(Ndate))

    Set THmax = min ((THmin + Rsm), THmax) // selection of minimum soil-
moisture to maintain crop growth.
    while(THmax>Csm) // condition for watering plants till required moisture level is
reached.
    {
    send 1 to relay to start irrigation
    }
       send 0 to relay to stop irrigation
    }
    else
        {
              send 0 to relay to stop irrigation
           }
       }
       else
       {
              Set the date to start irrigation
if( current date > = irrigation date)
{
while( THmax>Csm)
                 {
                    send 1 to start irrigation
                 }
             }
             Else
             {
                send 0 to stop irrigation
             }
    }
```

After calibrating the threshold limits, irrigation cycle is scheduled. For sectors classified into moist, irrigation is stopped as adequate amount of water is supplied. Watering these sectors will lead to over-flow of water that contradicts the aim of the proposed work. The wet classified sectors can be noticed when large quantities of water are absorbed by the soil, i.e., in situations like when a rain occurs. Thus, with schedulers at work, irrigation of sectors happens, making the system use optimum amount of water for cultivation of crops in large field areas (Fig. 11).

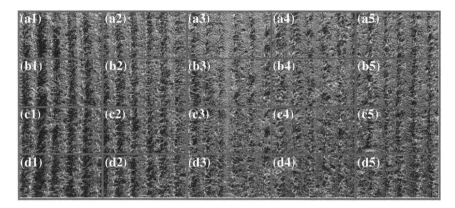

Fig. 11 Division of crop field into sectors

5 Experimental Results and Discussions

From the observation of field sensor data, sensor values are classified into classes based on soil moisture values recorded with values greater than 90 as most dry, values ranging from 81 to 90 as dry, values from 61 to 80 as moist and values from 40 to 60 as wet, respectively. The sensor data are then given as input to the models to predict future values. Output values thus derived are given as input to various classification algorithms, for validating the predicted values (Table 1).

The sensor values were evaluated with various classifiers, and the above results had been obtained. The results show high accuracy with more than 90% from classifiers Naive Bayes, classification via regression, multilayer perceptron, bagging and SMO with 88%, respectively, indicating that the values predicted from the regression are accurate enough for real-time consideration. From the results, it is evident that the proposed model can constantly check for readings from UV intensity, air temperature, soil temperature, soil moisture and start irrigation based on the readings.

Table 1 Field sensor data

UV	Air temp.	Air humidity	Soil temp.	Soil mois.
0.53	34	47	33.44	43
0.46	34	47	33.38	42
0.51	34	47	33.31	43
0.47	34	47	33.38	41
0.48	34	47	33.44	42
0.46	34	47	33.25	43
0.52	34	47	33.44	43
0.43	34	47	33.38	42
0.5	34	47	33.38	43

Table 2 Performance evaluation of proposed method with various classifiers

Classifier	Accuracy	Precision	Recall	F-measure
Bayes	92.95	0.90	1.00	0.94
MLP	95.97	0.99	1.00	0.99
Regression	99.49	1.00	1.00	1.00
Bagging	99.49	1.00	1.00	1.00
SMO	88.42	0.89	1.00	0.94

The proposed work can water the plant accordingly to maintain adequate quantities of water as required by the plant, automatically (Table 2).

It is also observed that there is a difference of 5 °C between soil temperature and air temperature. Key mechanism of the system does not let the temperature difference go beyond the determined level, so that the water content in the soil is always maintained at optimum levels.

6 Conclusion

The proposed system has been designed to take inputs from various field sensors that measure various factors that influence the growth of the plant. Considerable calibration has been done with the field values received from the plant specified, to prevent drying or over-flow of water such that the plant is supplied with ample quantity of water to survive with high yield. Relationship between soil moisture, soil temperature, air temperature, air humidity and UV intensity in the atmosphere has been carefully studied to evaluate a real-time plant irrigation model. The sensor data values are given as input to the SVR algorithm to predict future soil moisture levels, the predicted values are given to classifiers for accuracy, and among all the classifiers, classification via regression turns out to be with the highest accuracy, followed by bagging, MLP and Naive Bayes. Hence, the proposed model is successful in irrigating the plants with the values from field sensors and the model is successful in irrigating the plant through tough conditions and can get good yields. Further focus of the proposed work comes to the automation of the prediction of soil moisture, by using Web Services and Internet protocols. As with rapid technological developments, it is possible for the ML techniques to take over the human work force, hence, automation of the proposed system, along with development in weather forecasting systems, usage of weather data with forecast of weather from upcoming days, as a rainfall would be evident to schedule the irrigation based on rain forecast and save fresh water resources and avoiding over-watering the plant.

References

1. M. Cobaner, Evapotranspiration estimation by two different neuro-fuzzy inference systems. J. Hydrol. **398**, 292–302 (2011)
2. S.L. Davis, M.D. Dukes, G.L. Miller, Landscape irrigation by evapotranspiration-based irrigation controllers under dry conditions in Southwest Florida. Agric. Water Manag. **96**(12), 1828–1836 (2009)
3. S.L. Davis, M.D. Dukes, Irrigation scheduling performance by evapotranspiration-based controllers. Agric. Water Manag. **98**(1), 19–28 (2010)
4. H. Drucker, C.J. Burges, L. Kaufman, A.J. Smola, V. Vapnik, Support vector regression machines, in *Advances in Neural Information Processing Systems* (1997), pp. 155–161
5. M.K. Gill, T. Asefa, M.W. Kemblowski, M. McKee, Soil moisture prediction using support vector machines 1. JAWRA J. Am. Water Resour. Assoc. **42**(4), 1033–1046 (2006)
6. A. Goldstein, L. Fink, A. Meitin, S. Bohadana, O. Lutenberg, G. Ravid, Applying machine learning on sensor data for irrigation recommendations: revealing the agronomist's tacit knowledge. Precis. Agric. **19**(3), 421–444 (2018)
7. J. Gubbi, R. Buyya, S. Marusic, M. Palaniswami, Internet of Things (IoT): a vision, architectural elements, and future directions. Future Gener. Comput. Syst. **29**(7), 1645–1660 (2013)
8. T. Ojha, S. Misra, N.S. Raghuwanshi, Wireless sensor networks for agriculture: the state-of-the-art in practice and future challenges. Comput. Electron. Agric. **118**, 66–84 (2015)
9. F. Viani, M. Bertolli, M. Salucci, A. Polo, Low-cost wireless monitoring and decision support for water saving in agriculture. IEEE Sens. J. **17**(13), 4299–4309 (2017)
10. F. Viani, Experimental validation of a wireless system for the irrigation management in smart farming applications. Microwave Opt. Technol. Lett. **58**(9), 2186–2189 (2016)
11. T. Wark, P. Corke, P. Sikka, L. Klingbeil, Y. Guo, C. Crossman, G. Bishop-Hurley, Transforming agriculture through pervasive wireless sensor networks. IEEE Pervasive Comput. **6**(2), 50–57 (2007)

Optimized Cylindrical and Rectangular DR Antenna for Ultra-Wideband Applications

K. Srinivasa Naik, D. Madhusudhan, S. Chandini and S. Aruna

1 Introduction

On human lives, the impact of wireless communications has become a part of daily life. Compact and efficient radiators are needed to achieve better wireless communications applications. One of the best radiators in microwave frequencies is dielectric resonator antenna (DRA). In microwave frequency applications, DRA is a good component. To fabricate DRA with high dielectric constant low loss materials are required with different shapes and dimensions. By using several improvement methods, impedance bandwidths can be increased. To transmit high data rate WLAN to meet the UWB range, DRAs are designed, and also, they are showing better properties in radar and micro-imaging applications [1–7].

2 Design Considerations

In this work, a *T*-shaped fed rectangular DRA (RDRA) has been designed for UWB applications with broadband impedance bandwidth from 0.31 to 0.55 GHz which can be obtained to cover both IEEE 802.11a WLAN and BAN frequencies, The

K. Srinivasa Naik (✉) · D. Madhusudhan · S. Chandini
Vignan's Institute of Information Technology (A), Visakhapatnam, Andhra Pradesh, India
e-mail: nivas97033205@gmail.com

D. Madhusudhan
e-mail: madhudonga@gmail.com

S. Chandini
e-mail: chandinispavan@gmail.com

S. Aruna
Andhra University College of Engineering, Visakhapatnam, Andhra Pradesh, India
e-mail: aruna9490564519@gmail.com

overall dimensions of this antenna is 30.0×21.0 mm^2. For this design, lower UWB frequency band is selected, i.e., from 3.1 to 4.9 GHz with a dielectric constant of 9.4.

The designed antenna is shown in Fig. 1a, b. Rectangular ceramic block dimensions is $6.00 \times 9.00 \times 6.00$ mm^3. FR4 Substrate with thickness 0.8 mm with permittivity of 4.5 to realize 50 Ω feed line of 18 mm $\times$ 1.5 mm dimensions is printed on the substrate.

Asymmetrical DRA with wideband single slot-fed antenna is designed with a pair of adjacently grouped CDRAs. The designed antenna dimensions are 30×25 mm^2, FR4 substrate with $\varepsilon_r = 4.5$, $t = 0.8$ mm. The feed line dimensions are

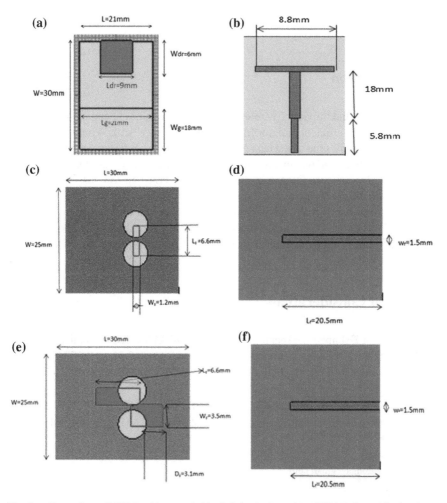

Fig. 1 a Front view of RDRA with ceramic block, b back view of modified T-shaped feed, c front view of CDRA, d back view of CDRA, e front view of cylindrical DRA for k_u band applications, f back view of cylindrical DRA for k_u band applications

20.5×1.5 mm^2 and symmetrically placed with respect to the coupling aperture. The proposed asymmetrical CDRA is shown in Fig. 1c, d.

A rectangular aperture feed asymmetrical two cylindrical dielectric resonators are designed for UWB applications. The impedance bandwidth covers dual-band frequency in between 6.02 to 7.32 GHz and 8.72 to 16.57 GHz with a gain of 8 dBi. The feed lengths are 20.5×1.5 mm^2.

3 Simulated Results

A. Return loss for RDRA

The S parameter versus frequency plot of RDRA is shown in Fig. 2, and it resonates at 3.5 GHz frequency with return loss of -17.2 dB.

The VSWR ratio at a frequency of 3.5 GHz is 1: 1.31, respectively, for rectangular DRA, which is shown in Fig. 3.

B. Cylindrical DRA parameters

The S parameters of a cylindrical DRA which resonate at a frequency of 12.2 GHz are shown in Fig. 4. The return loss is -49.2 dB.

The VSWR value at 12.2 GHz frequency is 1.31, respectively, for cylindrical DRA which is shown in Fig. 5.

C. Cylindrical DRA with asymmetric feed parameters

The return loss of CDRA with asymmetric feed which reonates at 13.44 GHz are shown Fig. 6, and the return loss is -31.1 dB.

The VSWR value at a frequency of 13.4 GHz is 1.31, respectively, for rectangular DRA using CST, which is shown in Fig. 7.

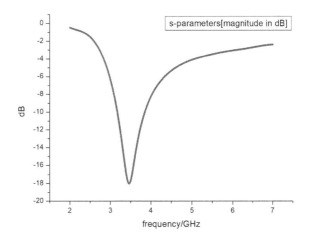

Fig. 2 S-parameter curve of a rectangular DRA for UWB applications

Fig. 3 VSWR of a rectangular DRA of UWB applications

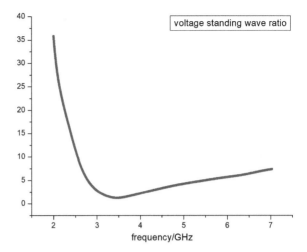

Fig. 4 *S*-parameter curve of a cylindrical DRA

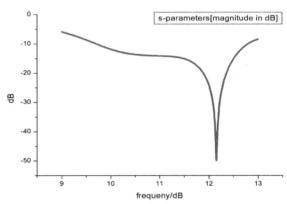

Fig. 5 VSWR of a cylindrical DRA

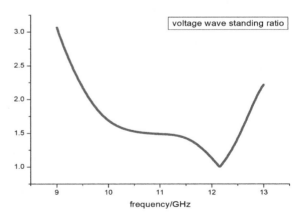

Fig. 6 Impedance bandwidth cylindrical of DRA using for k_u band applications

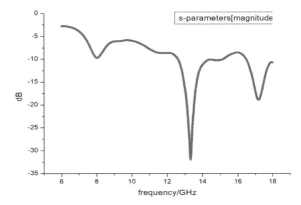

Fig. 7 VSWR of cylindrical DRA using for k_u band applications

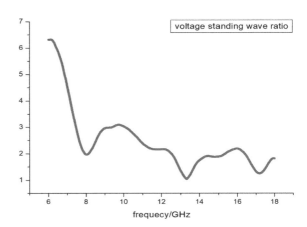

4 Conclusion

A lower UWB band rectangular DRA has been designed with micro-band power matching lines on FR4 substrate with average gain of 3 dB for WiMAX applications. The performance of the proposed antennas can be increased by using Hybrid DRAs with microstrip line feed. To achieve best spectrum performance, asymmetric DRAs are used and optimized. The total bandwidth achieved is 29% of frequency range 9.62–12.9 GHz, with gain of 8 dB. An asymmetric pair of cylindrical DRA with zigzag opening is designed. In this work, 62% of bandwidth is achieved which covers dual-band frequency range from 6.02 to 7.32 GHz and 8.72 to 16.57 GHz, and a gain of 9 dBi is obtained.

References

1. Y. Gao, Z. Feng, L. Zhang, Experimental investigation of new radiating mode in rectangular hybrid dielectric resonator antenna. IEEE Antennas Wirel. Propag. Lett. **10**, 91–94 (2011)
2. K.K. So, K.W. Leung, Bandwidth enhancement and frequency tuning of the dielectric resonator antenna using a parasitic slot in the ground plane. IEEE Trans. Antennas Propag. **53**, 4169–4172 (2005)
3. M. Abedian, S.K.A. Rahim, Sh Danesh, S. Hakimi, L.Y. Cheong, M.H. Jamaluddin, Novel design of compact UWB dielectric resonator antenna with dual band rejection characteristics for WiMAX/WLAN bands. IEEE Antennas Wirel. Propag. Lett. **14**, 245–248 (2015)
4. S. Danesh, S.K.A. Rahim, M. Abedian, M.R. Hamid, A compact frequency reconfigurable rectangular dielectric resonator antenna for LTE/WWAN and WLAN applications. IEEE Antennas Wirel. Propag. Lett. **14**, 486–489 (2015)
5. K. Srinivasa Naik et.al., Design, analysis and parametric study of rectangular dielectric resonator antenna arrays. Int. J. Adv. Sci. Technol. 112, 67–78 (2018)
6. K. Srinivasa Naik et.al., Design and analysis of different patch geometry and complementary split ring resonator for X-band applications. Int. J. Recent Technol. Eng. (IJRTE) **7**(5S4), 858–869 (2019)
7. Sh. Danesh, S.K.A. Rahim, M. Abedian, Frequency reconfigurable rectangular dielectric resonator antenna for WiMAX/WLAN applications. Microwave Opt. Technol. Lett. **57**(3), 579–582 (2015)

A Stitch in Time Saves Nine: A Big Data Analytics Perspective

T. Archana Acharya and P. Veda Upasan

1 Introduction

> A customer is the most important visitor on our premises. He is not dependent on us. We are dependent on him. He is not an interruption in our work – he is the purpose of it. We are not doing him a favour by serving him. He is doing us a favour by giving us the opportunity to serve him.
>
> Mahatma Gandhi

Today's digital era is throwing many avenues of opportunities but at the same time it is also creating many avenues of threats. Traditional data base management systems are providing transactional comforts like ATMs, credit/debit cards, etc. But in practicality the traditional database systems are unable to generate the reports within time to make an effective decision for a current problem as the processing and retrieval of such data are time-consuming, thus paralysing the total system. Digitalisation is generating complex, large-volume data every internet second. The different sources of data are Facebook, WhatsApp, YouTube, Twitter, GPS signals, mobile phones and many more. The industrial survey reports states that the generated data comprise 10% of structured data, 70% of unstructured data and 20% of semi-structured data. To get competitive advantage any business in general and banking industry in particular has to generate a mechanism to generate reports within time to address the issues of the customers to increase their satisfaction levels. The mechanism should be able to provide a platform to understand the customer behavioural patterns. The challenges

---

T. Archana Acharya (✉)
Department of Management Studies, Vignan's Institute of Information Technology, Beside VSEZ, Duvvada, Visakhapatnam 530049, India
e-mail: taamphil@gmail.com

P. Veda Upasan
Department of Computer Science and Systems Engineering, College of Engineering, Andhra University, Visakhapatnam 530003, India
e-mail: vedaupasan142@gmail.com

© Springer Nature Singapore Pte Ltd. 2020
J. Fiaidhi et al. (eds.), *Smart Technologies in Data Science and Communication*, Lecture Notes in Networks and Systems 105, https://doi.org/10.1007/978-981-15-2407-3_28

ahead are gigantic form of the flowing data which requires proper quantification and converting data sets into information for further application. Big data are the mode of mechanism for effective conversion.

1.1 Research Gap and Research Problem

In banking sector, voluminous, variety and complex data are available where in time requirement of processed data for effective decision making is imperative to meet the customer satisfaction. The traditional data base system is proved to be inefficient. The research problem identified is to reduce the processing time of data and make the information available at the right time for decision making so that the needs of customers are addressed at the earliest. The dynamic environment calls for adaptable solutions for increasing customer satisfaction.

1.2 Objectives of the Study

1. To Understand the concept of the new technology
2. To Review its role in financial institutions
3. To Evaluate the applications in determining customer satisfaction.

1.3 The Concept

What is data?

Data are collection of facts. Data can be divided into three categories (Fig. 1).

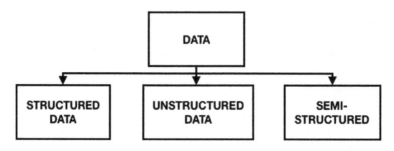

Fig. 1 Classification of data [1]

Structured data is based on defining a data model for which the fields are specified that is what type of data is required to be stored, processed and accessed. For example, a banking industry requires types of depositors, their transactional behaviour, frequency of transactions, volume of transactions, etc. The type of data is generally numeric, currency, alphabetic, name, date, address, etc. Therefore, structured data are data sets with defined set of fields in a particular file or record. Relational databases and spreadsheets contain this type of data. The key feature is that data can be easily stored, queried and analysed. The limitation is that any data which are not defined under any specified field either have to be recorded on paper or outside the field where the relation between the fields could not be established. Structured Query Language (SQL)—a programming language is defined to handle this type data under relational database management systems.

Unstructured data are a data type which generally cannot be defined under specific fields. This data type generally includes photographic images, photographs, videos, web pages, pdf files, power point presentations, blog entries, emails, etc.

Semi-structured data are the type of data which include specific fields and non-specific fields but it cannot be adhered to the data model structure. For example, emails have structured data like sender, recipient, date, time, etc., but the text message and attachment come under unstructured data.

What is big data?

Big data literally mean data which is big. The word big suggests massiveness, voluminous and complexity. Technically data which go beyond the capacity of storage are called big data. Generally, the data sets size is beyond terabytes or petabytes. The following figure shows the various sources of big data (Fig. 2).

The characteristics features are defined as 5 V's concept. The following figure depicts the 5 V's concept (Fig. 3).

Volume: There is large amount of data sets that are generated which is higher than terabytes–petabyte–exabyte. The main sources of generation are corresponded

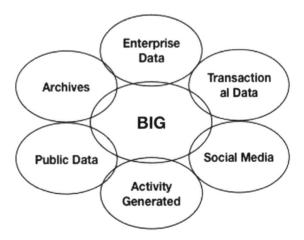

Fig. 2 Various sources of data [1]

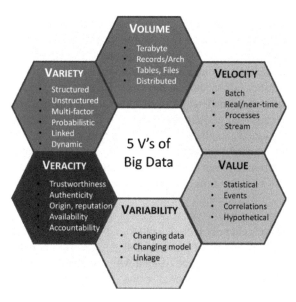

Fig. 3 Concept of 5 V's [2]

to digitalised transactions, through social media, etc. The data generation by 2025 is estimated to be around 175 zetabytes which is 300 times of 2005 data generation (Fig. 4).

Variety: The data sets are classified into structured, semi structured and unstructured. The variety is defined by different types of data sets. Out of the total data 80% of it correspond to unstructured data. The following chart shows different types of data sets generating from different sources (Fig. 5).

Velocity: The speed at which the data are generated is termed as velocity of data. The speed at which data are generated the speed of the processing is not matched from the traditional data base systems as it is sequential in nature; hence, there is requirement for parallel processing. As on date, the chart shown in Fig. 6 shows the different sources generating variety of data.

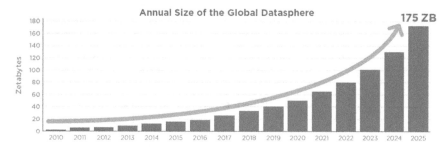

Fig. 4 Volume of data [3]

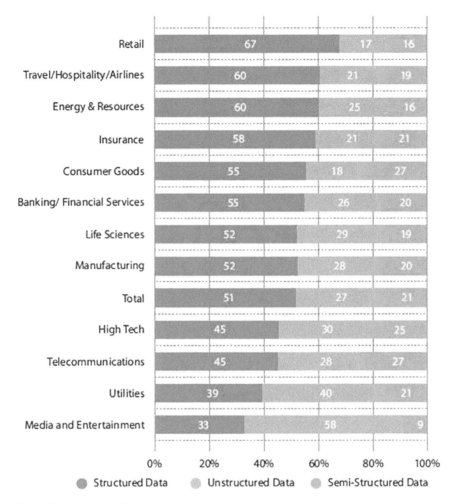

Fig. 5 Variety of data [4]

Different sources generating variety of data at great velocity includes social media like Facebook, Instagram etc. It is expected that the velocity of data generation by 2020 will reach 194.4 exabytes per month.

Veracity: Veracity refers to most important part of big data that is about reliability of such data sets. Reliability refers to quality and what percentage of uncertainty presents in data sets. It calls for trustworthiness. For decision making, trustworthiness is the base for which data quality and accuracy are the two components required.

Value: The value is a reference point created when generated data are accessed and utilised. A survey report reveals that only 0.5% of the data generated is accessed for utilisation. With volumes of data sets in hands, splendid value can be created with big data analytics.

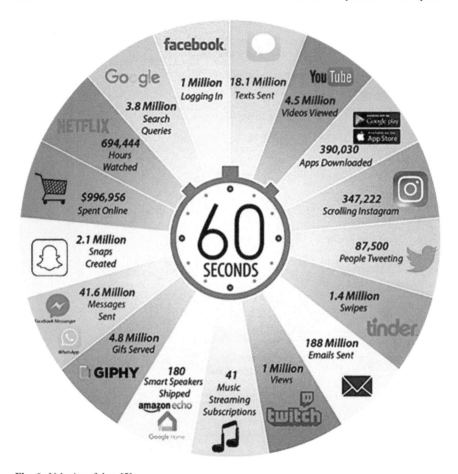

Fig. 6 Velocity of data [5]

What is Big Data Analytics?

Creation of value is achieved only when proper information is extracted from data sets for the given situation. To create value from volume, variety, velocity and veracity, there should be optimal processing power, capability of analytics and skill sets to arrange the information as per end-user requirements.

Collection of data, processing of data and management of data are called as data processing which results in generation of information as per the requirement of the end-user. Big data analysis have the following four steps called as 4 A's concept: acquisition, assembly, analyse and action. Acquisition means collection of data sets from various different sources. Assembly means identification and separation of data sets as per the fields. This is a time taking process as the volume, variety and velocity are the defining components of the time duration. Analyse helps in authenticating the reliability of the data sets which usually done by running queries, developing

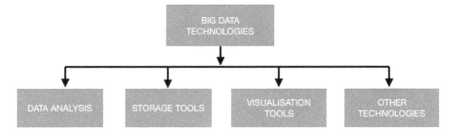

Fig. 7 Big data technologies [1]

models and building algorithms to enhance the data quality and accuracy. Action provides value to the process as it generates valuable knowledge for the end-user how to understand and interpret such knowledge.

The architecture consists of four layers: data layer, analytics layer, integration layer and decision layer. The data layer is based on RDBMS which is based on structured, unstructured and semi-structured data bases. For semi-structured and unstructured data base, NOSQL databases are used. HBase, Hive, Spark and Storm (software tools) are used to streamline the databases and to support the software tools Hadoop and MapReduce are used. In analytics layer, the performance of analytical engine is improved by developing environment using models which modify the data into regular intervals. Implementation of dynamic data analytics and real-time values are deployed with the help of this environment. The applications of end-user and analytical engine are integrated in the integration layer. Finally, the decision layers provide the platform for end product to market. Mobile apps, desktop applications, business intelligence software and interactive web applications are examples of end product.

The following chart depicts the big data technologies (Fig. 7).

The big data technologies are basically grouped under four different heads. Firstly, data analysis includes Hadoop, Map Reduce, Hive, PIG, WibiData, Platfora, Rapid Miner. Hadoop is an open source platform which is flexible and user friendly either to gather, access or process the data. For example, locational base data including weather, traffic sensors and social media data are some of the applications. For scalability of the data, a programming environment called as Map Reduce is used. There are two main functions: Map Task: converting data sets into different sets of value pairs and Reduce Task: reduced tubules are formed by combining various task of Map Task. The following figures show the input and output of Map Reduce (Fig. 8).

Hive is an open source software where the bridges permit predictable business applications to run SQL queries against a Hadoop cluster. PIG is a language which helps in the execution of query over data stored on a Hadoop (Fig. 9).

Wibidata helps to personalise customer experiences by allowing real-time responses to user like recommendations, decisions and serving personalised content. Lastly Rapid Miner is a software platform which is an integration of machine

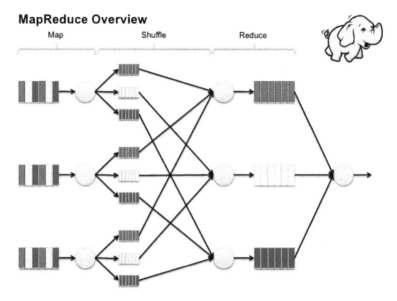

Fig. 8 Map reduce [6]

Fig. 9 Big data analytics tools [7]

learning, data mining, text mining, predictive analytics and business analytics. The growing data call for more storage space. Thus, in big data storage goes with data compression and storage visualisation. Data storage technologies include Hbase, Sky tree and NoSQL which work on non-relational databases. Visualisation tools includes R tool, Tableau, Infogram, ChartBlocks, Ember Charts and Tangle which help in presenting the information in an organised form as per the end-user. The

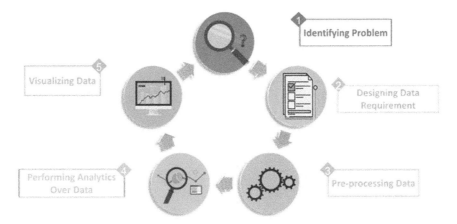

Fig. 10 Big data analytics process [8]

following figure displays the flow chart of big data analytics process. The steps include:

1. To identify the problem
2. To designing the data requirement
3. To pre-process data (cleaning data)
4. To perform analytics
5. To visualise data (Fig. 10).

The input data are in various formats like the structured data, unstructured data and semi-structured data. The big data is processed through the data analytics funnel which has three steps refinement, structuring and linking and finally synthesis and enhancement. In this step, voluminous data are decreased into organised information for the end-user. The benefits of the process are cost reduction, faster and better decisions, new products and services (Fig. 11).

1.4 Importance in Financial Institutions

The key business challenge for customer centric business services like banking services is to anticipate the needs of the customer by analysing their behaviour. Typically, the financial institutions have data warehouses and tools of business intelligence for reporting the analysis of customer behaviour to achieve optimised operations. But in time information it is the most imperative requirement for these financial institutions to keep their place in the market, thus calling for predictive capabilities. Big data analytics provides higher levels of insights into faster data, thus enabling effective decision making and thus optimising operations. The following are the challenges faced by financial institutions (Table 1).

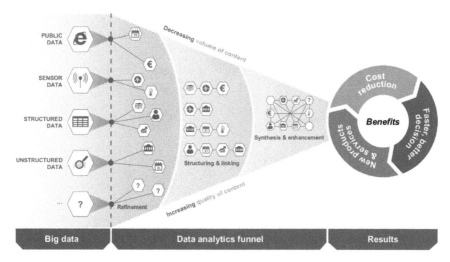

Fig. 11 Flow chart of data transformation [9]

Table 1 Challenges of financial institutions [10]

Functional area	Challenge
Increase revenue and improve profit margin	Increase productivity and profitability
Risk management	Reduce risk and cost
Fraud detection	To detect fraudulent transactions before in hand to reduce cost
Broker and trade compliance	To identify unfair trading activity
Anti-money laundering	To reduce exposure by detecting fraudulent transactions with speed and accuracy
Marketing and customer 360	Growth of customer base by increasing the products and services
Reputational risk	Protecting the brand

The challenges can be addressed effectively by using big data analytics. The following table lists how functional areas are solved by big data analytics (Table 2).

There are three types of analytics: prescriptive analytics, predictive analytics and preventive analytics. In financial services, predictive analytics is used. The following chart depicts the six business drivers for predictive analytics (Fig. 12).

1.5 Application of Big Data Analytics

The measurement of success for any business in general and banking sector in particular is profitability and productivity. To achieve success, customer is the base. So,

Table 2 Advantages of big data applications [11]

Functional area	Advantage of big data application
Fraud detection and prevention	No unauthorised transaction. Authenticated access given to systems so as to provide security and safety
Customer segmentation	Dividing customer base into groups based on age, gender, financial conditions, interests and spending habits with relevance to marketing and business to provide or deliver to customers with exactly what they are looking for
Management of risk, broker and trade compliance, anti-money laundering, reputational risk	Risk reduction and management of risk—reduces not only frauds in the transactions but also the chances of losing data
Marketing and customer 360, customer profitability	Understanding and identifying valuable customer helps banks to provide fiduciary services to its customers that is which investments give customers profitable returns
Analysis of past data and future predictions—increase revenue and improve profit margin	Prediction of future businesses based on data patterns. Understanding and spotting data patterns in various domains of banking services help banks in investment decisions: labour, money and time for profitable returns

Customer Insight
- Sentiment analysis
- Social media analysis
- Credit analysis
- Customer profitability and lifetime value

Customer Experience Management
- Virtual agent
- Next best action
- Relationship pricing

Channel Execution
- Next best offer
- Profitability
- Sales /channel performance

Business Strategy
- Product channel strategy
- Risk and capital modeling

Risk Management
- Risk and capital management
- Risk adjusted pricing
- Portfolio risk management
- Fraud/AML

Marketing
- Lead generation
- Customer segmentation
- Campaign management

Fig. 12 Business drivers of predictive analytics [12]

acquiring, developing, retaining customers become the prime motive of any organisation. The present technology addresses all the issues, thus defining its imperativeness in the current scenario. The following three important reasons call for application of big data analytics:

Table 3 Applications in banking sector [1]

Technology	Process	Outcome	Application in banks
Clustering	Within the data set: to identify meaningful data patterns and to find correlation	1. Segmentation: breaking data into simple parts. 2. To identify important data patterns	Frequency of financial services usage and potential of the customer forms the basis for **Customer Segmentation** Understanding the correlation between the demographic variables of the customers and segmentation to develop meaningful data patterns which aid in development of loyal targeted programs so that price optimising can be achieved especially in card usage
Analytics of text	Target fields are defined based on theory of probability. Target fields include unique features like rarity that form the key words and their occurrence. Target fields are used in predicting and defining the overall idea with classification algorithm	1. Generation of summary base documents compilation 2. Automatic assistance in reading the summary to identify target fields	In banks, the target fields identified can help in firstly distributing target fields into relevant channels, secondly to record and analyses behaviour of customers in different channels, thirdly to measure the effectiveness of particular channel, fourthly fiduciary function that is guidance to the customers towards favourable channels—**Channel Journey**
Neural networks	Activation of algorithm nodes which generate active signals to activate other nodes	1. Data assigned to predefined target field 2. Based on the assigning of data questions related to the event is answered resulting in action 1 or action 2	Customer's opinion is base to improve the quality of financial services. **Sentiment Analysis** provides an understanding about customer's opinion: monitoring the customer's opinions, to examine feedback. This understanding helps in identifying key customers there by improvising quality services through quality products

(continued)

Table 3 (continued)

Technology	Process	Outcome	Application in banks
Link analysis	To represent relationship between objects-based graph theory	1. Identify key links and its related key information 2. Analysing the links based on call patterns to find influential customers 3. To recruit new subscribers	**Revenue product bundling technique**: Helps in anticipating customer's financial goals to provide **Best offers**. This results in increasing loyalty of customers with increase in product propensity. To provide best offers banks find out retailers to offer customised product discounts and service discounts
Survival analysis: time to event analysis	Hazard probabilities and survival curves form the basis for evaluation of the critical events	1. Identify the dissatisfied customers who are leaving 2. Determining factors which increase/decrease the tenure of the customer 3. Understanding the movement of the customers from one segment to other segment by defining the various factors that contribute to time period the factors w time period of when is the customer likely to leave which factors likely increase or decrease customer tenure effects of various factors time period of when the customer moves to a new customer segment	**Understanding 360° of Customer** 1. Customers—profile and capability 2. Product engagement 3. Detecting leaving customers 4. Analysing potential loss of customers called as **Customer 360°**

(continued)

Table 3 (continued)

Technology	Process	Outcome	Application in banks
Decision trees	To handle diverse problem arrays, data mining techniques are used to handle various data types	1. Decreasing the volume of data 2. Splitting the data into small data cells	Protecting the customer's interest by Design Targeted marketed programs: Identifying the potential customers, detecting before occurrence of various types of frauds, alerting suspicious transactions, reporting and disclosure of mis-regulatory actions. This technology aids in **addressing security concerns and efficient management of risk**
Random trees	Identifying the difference between individual noise decision tree and errors possible	1. Identifying the gap in management of product	For service industry the customer is the base. Identifying the position of the customer in product lifecycle—helps in **product management**. Thus, aids in identifying the areas of the Product Lifecyle which requires marketing so as to impact the promotion of the bank products and services

1. Meeting needs of the customer daily by understanding their patterns of spending **improves satisfaction of the customers**
2. To increase profitability and overall revenues
3. To increase productivity.

The following table shows the application of big data technology in banks (Table 3).

The following table broadly offers the case studies in various international banks (Table 4).

In India, the implementation of analytics is still at infancy stage. Presently, the traditional data base system has been just incorporated with the digitalisation of the banking operations. The pace of the velocity of the data is demanding higher technological tools to be implemented immediately.

Table 4 Analytics in banks [1]

Bank	Functional area	Big Data tool	Purpose
Lloyds Banking Group	Fraud detection	Phone printing TM technology	To detect Fraud: 'Audio finger Print' is created using unique call features (around 1300 features analysed). The unique call features include type of call, noise of background, history of the number and call location. The uniqueness helps in identifying activities which are unusual like caller ID spoofing, distortion of voice, etc, which gives a base to detect potential frauds
Danske Bank	Increase revenue and improve profit margin	Advanced big data analytics	To evaluate behaviour of the customer and determine their preferences
Danske Bank	Fraud detection	Advanced analytics	To detect the fraud well in advance
Bank: JPMorgan Chase	All functional areas	COiN: contract intelligence	To extract important information from documents and to analyse them
European Bank	Customer 360	Machine-learning algorithms	To address shrinking customer base (15% rise in customer base)
US Bank	Reduction in costs	Machine-learning algorithms	To identify and correct unnecessary discounts of call patterns (8% rise in profitability)
Deutsche Bank	Reduction in costs	Multiple production Hadoop platforms	To process data at customer friendly cost
A top consumer bank in Asia	Customer segmentation	Advanced big data analytics	Next-product-to-buy model was build by creating data sets which include demographics of customers, key behaviours, customer preferred products, mobile and online transactions, information of credit-card usage, etc. This resulted in increasing the buying of products by three times

2 Conclusion

It can be concluded that today's Techology driven transactions are generating Data which is Voluminous and Complex including variety. For handling such data Big Data analytics is the effective and efficient tool. The analytics of big data helps the banking sector to handle its customer base more efficiently. Today customer is the emerging power of the market. With the impact of digital era day by day customer is becoming the deciding factor of productivity and profitability of the banks. High demand for faster and effective banking process, thus enhancing productivity, controlling risk and finally increasing profitability depend invariably on the way the data are handled, processed and reported. Thus, it becomes imperative for the banking sector to start its big journey in the world the technology.

References

1. Research work
2. https://www.researchgate.net/figure/The-five-Vs-of-Big-Data-Adapted-from-IBM-big-data-platform-Bringing-big-data-to-the_fig1_281404634
3. https://www.google.com/url?sa=i&rct=j&q=&esrc=s&source=images&cd=&cad=rja&uact=8&ved=2ahUKEwjNreGfk8nmAhVTzDgGHdQGCusQjRx6BAgBEAQ, https://www.datanami.com%2F2018%2F11%2F27%2Fglobal-datasphere-to-hit-175-zettabytes-by-2025-idc-says%2F&psig=AOvVaw1Iq_JD_fqIbXiDWwpfHJcv&ust=1577100473601967
4. https://www.google.com/url?sa=i&rct=j&q=&esrc=s&source=images&cd=&ved=2ahUKEwir7Ku_k8nmAhXcyzgGHYCgAhsQjRx6BAgBEAQ, https://sites.tcs.com%2Fbig-data-study%2Findustries-unstructured-data%2F&psig=AOvVaw0EsxebpMCMAhnxY9OEMBK4&ust=1577100538011242
5. https://www.google.com/url?sa=i&rct=j&q=&esrc=s&source=images&cd=&ved=2ahUKEwjpytKFlMnmAhX0jeYKHdcFAFQQjRx6BAgBEAQ, https://www.reddit.com%2Fr%2Fcoolguides%2Fcomments%2Fbcrwuy%2F2019_this_is_what_happens_in_an_internet_minute%2F&psig=AOvVaw3uLC_Ya9jaEOIQBOoRqs4J&ust=1577100680508874
6. https://www.google.com/url?sa=i&rct=j&q=&esrc=s&source=images&cd=&ved=2ahUKEwixufXClcnmAhW5yDgGHZTTBvoQjRx6BAgBEAQ, https://www.slideshare.net%2Fnarayan26%2Fdata-sciencewebinar-061312-13372205%2F41-MapReduce_Overview_Map_Shuffle_Reduce&psig=AOvVaw3tEXgMHQfCZA2RBPcE_SzA&ust=1577101084812363
7. https://www.google.com/url?sa=i&rct=j&q=&esrc=s&source=images&cd=&ved=2ahUKEwjGk9mqlMnmAhXI9nMBHRmxBRUQjRx6BAgBEAQ, https://www.houseofbots.com%2Fnews-detail%2F12023-1-see-some-best-known-big-data-tools%2C-there-advantages-and-disadvantages-to-analyze-your-data&psig=AOvVaw3rP4RZqxQ4brx6ztcxlBHY&ust=1577100765472507
8. https://www.google.com/url?sa=i&rct=j&q=&esrc=s&source=images&cd=&ved=2ahUKEwifx4GtlsnmAhUl83MBHXFdCiMQjRx6BAgBEAQ, https://www.quora.com%2FWhat-are-big-data-analytics&psig=AOvVaw0Qncdk2qjseSGvA_IiURHy&ust=1577101289836492

9. https://www.google.com/url?sa=i&rct=j&q=&esrc=s&source=images&cd=&ved=2ahUKEwjkltbjlsnmAhWf6XMBHQ8CD2gQjRx6BAgBEAQ, http://bvijtech.com%2Fanalytics%2F&psig=AOvVaw1eLkATzuG0VRta_W2-YW31&ust=1577101415244276
10. http://www.oracle.com/us/technologies/big-data/big-data-in-financial-services-wp-2415760.pdf
11. https://www.aspiresys.com/WhitePapers/bigdata-analytics-to-bank-on-your-biggest-asset-information.pdf
12. https://thefinancialbrand.com/46320/big-data-advanced-analytics-banking/

Flexi-Lexicon Learning Using Krill Herd Algorithm for Sentiment Analysis

Muddada Murali Krishna, JayaVani VanKara, V. Satyanarayana Kalahasthi and Ming Chen

1 Introduction

Opinion mining from online streaming media is a hectic task with highly challengeable forums such as analyzing the theme, classification of subjective opinions, emotions and sentimental contents on a particular object, or on an organization or opinion on a political statement, etc. [1]. In the recent trend, the sentiment analysis made it easy to work on such frameworks [2–5]. Lexicon and Corpus are the two base algorithms for effective sentiment classification from online stream data. A lexicon analysis defines a text in terms of positive and negative and addresses the certain level of scores for each. The scores indicate the different types of emotions or conclusions if they fall under the respective ranges [6–10].

Hu et al. [11] imposed the level of network data to produce emotional spread for analysis of sentiments in the obtained data. In [12], semantic level of models is incorporated for meaningful extraction from the stream data. Another approach in [13] utilizes the meta-data of a particular data for effective discrimination of sentinel data. In [8] from the stated vocabulary words from the repository, an add-on

---

M. M. Krishna (✉)
M.C.A Department, Dr. Lankapalli Bullayya PG College, Visakhapatnam, Andhra Pradesh, India
e-mail: muralikrishna1926@gmail.com

J. VanKara
Computer Science and Engineering, Dr. Lankapalli Bullayya College of Engineering for Women, Visakhapatnam, Andhra Pradesh, India
e-mail: jayavani.vankara@gmail.com

V. Satyanarayana Kalahasthi
Department of Computer Science and Engineering, Avanthi Institute of Engineering and Technology, Visakhapatnam, India
e-mail: thanqji@gmail.com

M. Chen
Harbin University of Commerce, Harbin, China
e-mail: mingchen@163.com

© Springer Nature Singapore Pte Ltd. 2020
J. Fiaidhi et al. (eds.), *Smart Technologies in Data Science and Communication*, Lecture Notes in Networks and Systems 105, https://doi.org/10.1007/978-981-15-2407-3_29

which counts the scores of each lexicon has been proposed. Saif et al. [11] proposed a different method to capture the sentiments from different perspectives. Coelletta et al. [12] proposed SVM classification which has been used for effectively classifying the different clauses of sentiments in the selected data. Lu in [10, 13] utilized the concept of identifying text similarities for effective sentiment classification. The limitations exist in these concepts intuited the researcher to propose a concept of finding polarity between the semantic content using Krill Herd algorithm.

In this paper, a method to identify the polarity using sentiment analysis in the online streaming data Twitter is proposed. An optimal level of dynamic lexicons is identified, and the problem is represented in the form of optimization problem that can be solved using evolutionary algorithms. Conventional KH algorithm has been used to tackle this problem, and the results are demonstrated with a confidence level interval of 0.95.

2 Problem Formulation

The initial process is to extract sentiment lexicon from the given streamed data for training phases.

Each lexicon will be assigned with a score within a range [min, max].

2.1 Sentiment Word Extraction

Words with a higher range of scores (i.e., more than 2) are considered as sentiment words from the given database. When a repeated word comes in the phrase, the algorithm fails to find the exact score for it since the meaning of that word cannot be predicted.

2.2 Optimization Problem

The theme of this research work is to find an optimal lexicon that minimizes the classification error. A method will be represented to identify the optimal lexicon namely Flexi-Lexicon (FL) which can be defined as

$$\text{FL}(D_m, T_i, L_k) = \sum_{w_j \in T_i} V_k(W_j)$$

where W_j represents the words with score more than 2, and $V_k(W_j)$ is the score of sentiment word W_j.

2.3 Krill Herd Algorithm

Krill Herd algorithm has the potential to search for an optimal solution in the given stamp of time. It is inspired from Krill's concept. There are three processes, which hold the search process in krills.

(i) Movements imposed by other krills
(ii) Foraging
(iii) Random diffusion.

And the combined solution can be represented as

$$\frac{dX_i}{dt} = N_i + F_i + D_i$$

where N represents the movement intuited by other krills, F mentions the foraging activity and D represents the diffusion.

The movement induced by other krills can be represented as

$$Ninew = N\,max\alpha i + \omega n Niold$$

And the local search can be represented as

$$\alpha i = \alpha ilocal + \alpha itarget$$

The foraging activity of an individual krill can be represented as

$$Fi = Vf\beta i + \omega f Fiold$$

And it is further added to the existing position as

$$\beta i = \beta ifood + \beta ibe$$

Physical diffusion comes with a random walk, and it can be represented as

$$D_i = D^{max}\left(1 - \frac{I}{I_{max}}\right)\delta$$

3 Experimental Results

The results of the proposed algorithm are tabulated in Table 1. The proposed algorithm has been tested in six different datasets which include Sanders, OMD, strict

Table 1 Performance of KH sent with other existing literature algorithms for six different datasets

Method	Overall accuracy (%)	Positive class			Negative class			Overall average F1 (%)
		Precision (%)	Recall (%)	F1 (%)	Precision (%)	Recall (%)	F1 (%)	
SOMD dataset								
Bing Liu Lexicon	68.66	70.46	30.21	37.89	69.42	92.49	78.84	58.38
Random search	52.70	38.74	45.54	41.61	64.21	57.13	59.75	50.41
ALGA	83.26	77.08	77.68	76.83	87.13	86.37	86.25	82.16
ALGA-SW	79.89	73.64	72.72	73.14	84.64	84.76	84.35	78.28
ALGA + 1, 2, 3 g	86.06	82.43	79.88	80.91	88.43	90.52	89.52	85.30
ALGA + lex	87.31	82.99	80.81	81.64	89.02	90.56	89.43	86.03
KH-sent	87.54	87.14	82.34	83.00	90.09	90.14	89.97	86.32
Sanders dataset								
Bing Liu Lexicon	73.12	73.97	71.35	71.82	76.87	74.04	69.86	70.40
Random search	51.82	47.68	42.15	44.30	54.90	60.22	56.53	50.11
ALGA	82.58	77.43	87.43	82.28	87.35	78.36	82.69	82.02
ALGA-SW	80.06	80.11	83.26	81.24	80.33	76.98	78.08	79.87
ALGA + 1, 2, 3 g	86.45	83.90	86.85	84.66	89.22	84.89	87.17	86.19
ALGA + lex	85.81	84.27	86.03	84.71	88.22	86.27	86.72	86.21
KH-sent	86.78	85.26	87.90	85.92	89.92	87.09	87.56	86.71
HCR dataset								
Bing Liu Lexicon	81.58	0.00	0.00	N/A	81.32	99.66	90.47	N/A
Random search	59.96	27.53	40.29	31.78	78.67	65.60	71.37	50.92
ALGA	76.42	37.79	32.83	34.87	84.90	87.31	85.82	60.03
ALGA-SW	73.82	38.39	34.11	35.60	81.51	84.34	83.02	59.54
ALGA +1, 2, 3 g	84.07	61.07	36.74	45.94	86.80	95.16	91.00	68.23
ALGA + lex	77.87	42.67	37.72	39.76	86.10	88.40	87.31	63.93
KH-sent	84.12	61.85	39.64	52.25	87.02	95.71	90.83	72.18
OMD dataset								

(continued)

Table 1 (continued)

Method	Overall accuracy (%)	Positive class			Negative class			Overall average F1 (%)
		Precision (%)	Recall (%)	F1 (%)	Precision (%)	Recall (%)	F1 (%)	
Bing Liu Lexicon	68.30	63.82	23.55	32.46	68.47	93.55	78.72	55.38
Random search	55.28	64.48	61.97	62.28	39.59	40.74	39.88	50.55
ALGA	79.70	74.41	69.75	72.27	83.65	85.96	84.33	78.35
ALGA-SW	76.38	68.28	65.92	67.04	80.22	82.35	81.13	73.82
ALGA + 1, 2, 3 g	81.27	76.23	71.14	73.87	83.77	87.03	85.82	79.79
ALGA + lex	80.23	74.99	69.81	71.69	83.18	86.84	84.64	78.19
KH-sent	82.83	77.64	72.99	74.94	85.13	88.22	86.60	80.40
STS dataset								
Bing Liu Lexicon	74.85	70.59	88.96	78.23	83.06	59.72	68.52	73.42
Random search	53.42	52.37	49.50	49.77	51.93	54.97	52.40	51.71
ALGA	77.90	78.32	77.25	77.67	77.11	79.48	78.06	78.05
ALGA-SW	80.47	82.29	77.86	79.94	78.46	83.45	80.29	79.99
ALGA + 1, 2, 3 g	83.28	80.30	88.05	83.08	86.26	78.80	81.58	82.58
ALGA + lex	81.48	81.61	80.55	80.49	80.94	82.77	81.40	80.84
KH-sent	85.94	84.20	89.66	86.19	88.25	83.48	85.36	86.07
SemEval dataset								
Bing Liu Lexicon	74.04	73.85	98.73	84.62	74.06	18.47	27.85	56.48
Random search	58.55	75.50	63.49	68.45	32.00	45.02	37.39	52.71
ALGA	77.77	82.73	86.35	84.49	64.07	56.96	60.38	72.50
ALGA-SW	74.31	80.58	84.43	82.29	57.45	50.95	54.00	68.35
ALGA + 1, 2, 3 g	81.43	85.12	89.41	87.01	70.54	85.66	66.86	76.98
ALGA + lex	79.73	83.92	88.18	86.42	67.66	61.17	63.98	74.68
KH-sent	85.77	89.56	89.64	90.71	7421.37	70.07	73.16	81.26

OMD, HCR, SemEval, and Stanford Dataset. The algorithm has been implemented in MATLAB 2016, and for comparison, standard literature-based algorithms are used.

From the results tabulated in Table 1, it can be concluded that the proposed KH algorithm for sentiment analysis outperforms other existing solutions in all the cases.

4 Conclusion

The proposed method FL represents the lexicon as an optimization problem, and the Krill-inspired algorithm works on this optimization problem and shows that the proposed algorithm can find the optimal lexicon within limited number of iterations. The proposed algorithm has been evaluated in six different datasets, and each shows the significance of the proposed algorithm with a confidence interval of 0.95.

References

1. B. Liu, *Sentiment Analysis and Opinion Mining* (Morgan & Claypool Publishers, 2012)
2. E. Fersini, E. Messina, F.A. Pozzi, Sentiment analysis: Bayesian ensemble learning. Decis. Supp. Syst. **68**, 26–38
3. R. Feldman, Techniques and applications for sentiment analysis. Commun. ACM **56**(4), 82–89 (2013)
4. B. Liu, Sentiment analysis: a multi-faceted problem. IEEE Intell. Syst. **25**(3), 76–80 (2010)
5. A. Ritter, S. Clark, O. Etzioni, in *Named entity recognition in tweets: an experimental study*. Proceedings of the Conference on Empirical Methods in Natural Language Processing, EMNLP'11, Association for Computational Linguistics, Stroudsburg, PA, USA, pp. 1524–1534 (2011)
6. P.D. Turney, in *Thumbs up or thumbs down?: semantic orientation applied to unsupervised classification of reviews*. Proceedings of the 40th Annual Meeting on Association for Computational Linguistics, ACL 02, Association for Computational Linguistics, Stroudsburg, USA, pp. 417–424 (2002)
7. K Karan Raj, E-learning platform development using advanced model driven methodology. Int. J. Private Cloud Comput. Environ. Manag. **3**(2), 13–20 (2016)
8. C. Choi, M. Park, G. Jung, J. Cha, Detecting menu in rendered web document using machine learning techniques. Int. J. Artif. Intell. Appl. Smart Devices **5**(1), 1–6 (2017)
9. Y. Yang, R. Li, L. Feng, Wheat varieties identification research based on sparse representation method of dictionary learning. Int. J. Hybrid Inf. Technol. **11**(2), 13–24 (2018)
10. D. Khaturia, Aditisaxena, S.M. Basha, N.Ch.S.N. Iyengar, R.D. Caytiles, A comparative study on airline recommendation system using sentimental analysis on customer tweets. Int. J. Adv. Sci. Technol. **111**, 107–114 (2018)
11. M. Alrefai, H. Faris, I. Aljarah, Sentiment analysis for arabic language: a brief survey of approaches and techniques. Int. J. Adv. Sci. Technol. **119**, 13–24 (2018)
12. O. Harfoushi, D. Hasan, Amazon machine learning versus microsoft azure machine learning as platforms for sentiment analysis. Int. J. Adv. Sci. Technol. **118**, 131–142 (2018)
13. D. Zeng, Y. Dai, J. Wang, F. Li, A.K. Sangaiah, Aspect based sentiment analysis by a linguistically regularized CNN with gated mechanism. J. Intell. Fuzzy Syst. **36**(5), 3971–3980 (2019)

Analysis of Queuing Model-Based Cloud Data Centers

K. V. Satyanarayana, K. Sudha, N. Thirupathi Rao and Ming Chen

1 Introduction

Presently, it is seen that new trends related to computer technology emerge on a daily basis, and one of those new trends is cloud computing, which is anticipated to bring an enormous change in the way one uses computer and Internet. A total software and its related applications required environment was being provided under cloud computing. The primary uses of cloud computing are the cloud service in terms of data storage and Web applications. Cloud service development tools by Amazon, Google App Engine and IBM, etc., are well accepted and utilized.

An important component of cloud computing is Infrastructure-as-a-Service (*IaaS*) [1, 2], which is the ability to remotely access computing resources. The remote access of the services provided by the cloud computing environment was the access to the network, services related to the routing and the storage-related issues and their applications. The major work and the application of the IaaS provider will be in supplying the basic services and the services related to the hardware and other services related

---

K. V. Satyanarayana (✉)
Department of Computer Science and Engineering, Avanthi Institute of Engineering and Technology, Visakhapatnam, India
e-mail: thanqji@gmail.com

K. Sudha
Department of Computer Science and Engineering, Lenora College of Engineering and Technology, Rajahmundry, India
e-mail: sudhakurapati04@gmail.com

N. T. Rao
Department of Computer Science and Engineering, Vignan's Institute of Information Technology (A), Visakhapatnam, AP 530049, India
e-mail: nakkathiru@gmail.com

M. Chen
Harbin University of Commerce, Harbin, China
e-mail: mingchen@163.com

to the administrative services which were required to accumulate several list of applications and a stage for the management of several set of applications which requires the services from the cloud and its related areas. Typical examples of *IaaS* include computer cycles, servers, storage, network and backup, etc. [3, 4]. One of the advantages of *IaaS* is that one can access very expensive data center resources through a rental means. The most and the very important point to be considered was that the cloud computing can be taken as the important aspect for both the providers of the cloud and the customers of the cloud computing [5].

2 Related Work

Extensive survey on cloud computing was highlighted by K. Jayapriya et al. [4] where a widespread survey on quality of service with respect to their implementation details, strong points and limitations were presented. A. Tripathi [6] provided a practical guide to queuing analysis and also reviewed some elementary concepts in probability and statistics. Presentation points of cloud computing information stations using [(*M/G/*1): (inf/GD model)] queuing system were arrived in [3] particularly for mean number of tasks and waiting time in the system.

Queuing models are very useful in the working and processing certain models and the data related to several applications like the intend of manufacture, shipping, transport and stocking systems in terms of capacities and control. Fundamental concepts in queuing theory are well documented in [7]. Both finite source and infinite source queuing models are dealt in the above reference. T V. Mathew [8] surveyed on a variety of mechanism and models those are used for studying and identifying the data center and its requirements and the performance, evaluation for providing the quality of service in IaaS cloud computing, and it is related using systems. Transient probabilities of *M/M/*1 queue were calculated by H. Wang [9] based on task time, number of customers and traffic intensity. W. Stallings [10] brought out an on the whole standpoint on cloud estimation criteria and tinted it with assist of simulation.

H. Gronewelt [11] proposed and analyzed a set of new queuing patterns which could be encountered while working on these models, and the classification part of these patterns was also discussed. A. Tripathi [6] verified and proved the *M/M/c* model for two servers which increase the performance of the network model over using one server by tumbling the length of the queue and waiting time. J.Sztrik [12] measured an *M/G/*1-like system in which the service time distribution was being presented by a Coxian series of several memory less stages. A new and novel approach based on conditional probabilities is used to obtain solution of such systems.

An exhibition of academic outcome for the $M/M/c/c + r$ queuing model with eager clients is offered by Chandrakala et al. [2] with mathematical design as a fundamental form of call centers including derivations of combined sharing for the waiting time and probability of customer service and blocking. K. Jayapriya [4] studied the *M/M/c* queue model with client desertion which was also treated as the Erlang-A model, which was sovereign and identically dispersed client discard period with an exponential distribution. Five statistical software packages for queuing theory were compared by K. Jayapriya [4].

3 Queuing System—Overview

Queuing theory is the subject of mathematics and its applied areas like the applied mathematics and the subject of statistics. It mainly deals with the waiting lines. It is tremendously functional in predicting, identifying and evaluating the performance of the system. Operations research is one of the applicable and mostly used subjects and subject strategy of the queuing systems. Customary queuing theory problems that were being observed by several users and the researchers are the customers going and visiting a store, corresponding to requirements incoming at a device. Queuing theory provides an extended period of regular values. It is not possible or being considered in the queuing system procedure to identify or observe the occurrence of whether the next event will occur or not occur. In queuing models or the queuing theory, the arrival times are to be considered as the random, and similarly, the service times are also random in nature. A queuing system (Fig. 1) can be mentioned as "the customers resolve to appear for a specified check, wait if the check cannot begin right away and go away following being offered." The term "customer" can be men, products, machines, etc.

The characterization of the queuing model or the queuing system was very useful and very important for processing the several systems or several applications using these queuing systems. These systems can be characterized with several features or the factors like arrival processes of the customers, the time taken for providing the service, the discipline of the service, the capacity of the service and the number of service stages involved in finalizing the completion of the service. The following is a standard notation system (Kendall's notation) of queuing systems $T/X/C/K/P/Z$ with:

T Probability distribution of inter-arrival times
X Probability distribution of service times
C Number of servers
K Capacity of the queue

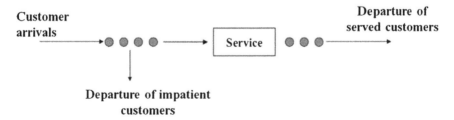

Fig. 1 Queuing model

P   Size of the population
Z   Discipline of the service.

Arrivals might initiate as of single or several sources referred to as the population those were being called. The population that was being called might be either limited or "unlimited." The arrival process of the system comprises of explaining how the clients or the users turn up to the system which was denoted by λ (inter-arrival rate). The service mechanism of a queuing system is precise in terms of number of servers (denoted by C) in which each server comprises of its possess line or an ordinary line and the probability sharing of clients check moment denoted by $1/\mu$. Here, "T" and "X" can take the following values:

M   Markovian (i.e., exponential)
G   General distribution
D   Deterministic
Er Erlang distribution.

Queue capacity (K) also denotes the loss of customers if queue is full. The size of the population (P) can be either finite or infinite. Service discipline (Z) basically can take the following values:

FCFS or FIFO   First come first served
LCFS or LIFO   Last come first served
RANDOM   Service in random order
GD   General discipline.

The discipline of a queuing system explains in detail the process of whether a system follows a rule or a regulation to a server that how the server identifies or selects the next customer or the next item from the existing queue or the queue under process from which the server completes the task given by a customer or an user.

4 Queuing Models and Formulations

Queuing models (Tables 2 and 3) are very much helpful to the users in predicting or identifying the performance of the service systems whenever there is a chance of existence of uncertainty in arrivals and service times to the system. The simplest possible (single stage, Fig. 1) queuing systems have the following components: customers, servers and a waiting area (queue). An arriving customer is placed in the queue until a server is available. To model such a system, we need to specify the characteristics of the arrival and service process; how (in what order) waiting customers are dispatched to available servers. For the present work, it is assumed that clients are offered the services in which order they arrive in the system (first come first served or *FCFS*). Mean value approach is used to determine the mean performance measures, *LS* and *WS,* directly by using Little's queuing formula and PASTA property.

Nomenclature for system parameters, chosen input values (S. No. 1–10) for the queuing models considered are highlighted at Table 1, and throughout, it is assumed that the system is in "steady state," i.e., it has operated for a long time with the same values for all the parameters.

The various formulations for the chosen queuing models are given at Tables 2 and 3.

Table 1 System parameters and performance measures

S. no	Description	Symbol	Inputs
1	Arrival rate in middle	λ	0–4
2	Service rates	μ	1, 3
3	Intensity of the traffic	$\in$	0, 1
4	Servers in number	C	1, 2
5	Customers number in size	K	4
6	Capacity of the buffers	R	4
7	Erlang parameter	er	2
8	Mean number of customers in a model	L_s	–
9	Total number of customers in queue	L_q	–
10	Waiting time of the customers in the queue	W_c	

Table 2 Formulations single-server models

Queuing model	L_s	L_q	W_s	W_q
M/G/1	$P + \frac{AP^2}{1-P}$	$\frac{AP^2}{1-P}$	$\frac{1}{\mu} + \frac{AP}{\mu(1-P)}$	$\frac{AP}{\mu(1-P)}$
M/Er/1	$P + \frac{\left(1+\frac{1}{er}\right)P^2}{2(1-P)}$	$P + \frac{\left(1+\frac{1}{er}\right)P^2}{2(1-P)}$	$\frac{1}{\mu} + \frac{\left(1+\frac{1}{er}\right)P}{2\mu(1-P)}$	$\frac{1}{\mu} + \frac{\left(1+\frac{1}{er}\right)P}{2\mu(1-P)}$

Table 3 Formulations multi-server models

Queuing model	M/M/c	M/M/c/c
L_s	$L_0 + P$	$\frac{\lambda . P0}{\mu}$
L_q	$\lambda . P + \frac{P(N>1)}{O(1-\theta)}$	0
W_s	$Wq + \frac{1}{\mu}$	$\frac{1}{\mu}$
W_q	$\lambda . P + \frac{P(N>1)}{M(1-\theta)}$	0
P_o	$P + \frac{\left(1+\frac{1}{er}\right)P^2}{2(1-P)}$	$P + \frac{\left(1+\frac{1}{er}\right)P^2}{2P}$

5 Results and Discussion

Evaluation of performance parameters was carried out by programming in MATLAB® 7.60 (R2008a) environment developed by MathWorks, Inc., USA. The input to the program is according to Table 1, and the output results for performance parameters are given at Figs. 2, 3, 4, 5.

Performance evaluation for single servers indicates that as the service rate (μ) increases for a constant range of traffic intensity (ρ) only waiting times of customers in the system (W_S) and queue (W_Q) decreases, whereas the length of customers in system (L_S) and queue (L_Q) remains unchanged as it is independent on μ. For the same input parameters, $M/D/1$ model shows optimum performance in terms of queue lengths and waiting times followed by $M/Er/1$, $M/M/1$. Performance of $M/G/1$ shows detrimental nature when compared with other queuing models, which is attributed to higher value of *CoV*. For higher-order Erlang parameters, $M/G/1$ and $M/Er/1$ models behave in close comparison.

Multiple server ($c = 4$) performance evaluation indicates that as the service rate (μ) increases for a constant range of traffic intensity (ρ), similar nature as in the case of single servers is observed with regard to queue lengths and waiting times. For the same input parameters, $M/M/c/c$ possesses least waiting times and customer lengths in queue, followed by $M/M/c/K$ ($K = 8$) and $M/M/c$. Performance of $M/M/c/c + r$ ($r = 8$) is detrimental in terms of waiting times in queue, whereas $M/M/c$ has higher customer queue lengths. Waiting times in queue for $M/M/c/c + r$ can be reduced by reducing the waiting capacity of customers in queue. Also, the performance with $M/M/c/K$ can be improved in terms of waiting times in queue by decreasing the maximum number of customers allowed in the system. Multiple servers always behave better in satisfying a queue than single servers.

6 Conclusions

Performance evaluation of small cloud computing data center is discussed with the theory based on queuing systems. Single server and multiple server models are presented along with their formulations for performance parameters. In order to analyze the performance of various network and queuing models, it is always not possible to analyze in a detailed way by normal calculations. Hence, an attempt had been made to analyze the model with the help of more number of numerical and graphical results by considering the mathematical models. Various mathematical models are available in the literature and discussed in detail for the better outputs and other results. Various equations and metrics were designed and considered for the better understanding of the performance of the considered model. All these models and methods were discussed and developed with the help of the mathematical models and solved them by analyzing and considering the code written in MATLAB software, and the results gave us more encouraging for better, and more experimentation is required further

Fig. 2 Single-server performance parameters, $\mu = 4$

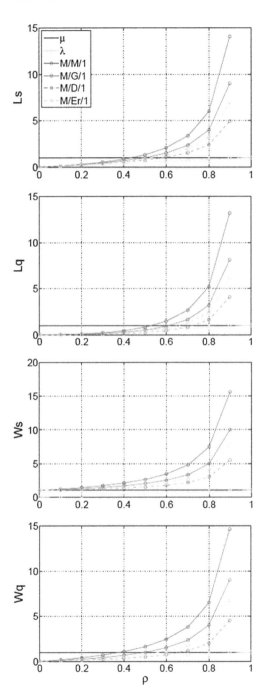

Fig. 3 Single-server performance parameters, $\mu = 8$

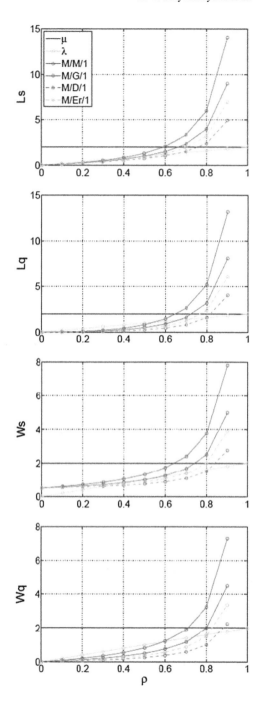

Fig. 4 Multi-server performance parameters, $\mu = 4$

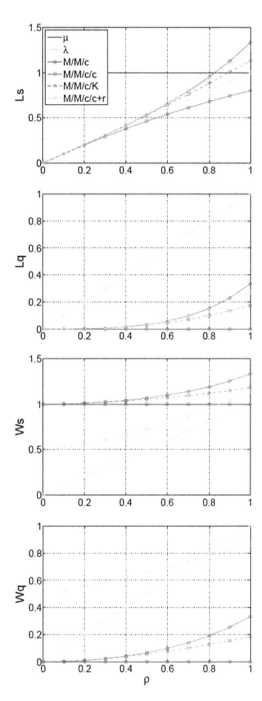

Fig. 5 Multi-server performance parameters, $\mu = 8$

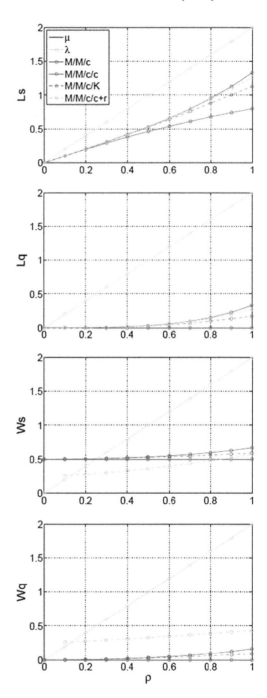

for better results with more cases of study. MATLAB programming/code generation and implementation for performance evaluation of cloud computing data server farm are accomplished. Comparisons among various models are attempted and relevant observations are highlighted.

References

1. I. Adan, J. Resing, *Queuing systems* (Eindhoven University of Technology, The Netherlands, 2015)
2. Chandrakala, et al., Survey on models to investigate data center performance and QoS in cloud computing infrastructure, in *First International Conference on Recent Advances in Science & Engineering* (2014)
3. K. Saravanan, et al., Performance factors of cloud computing data centers using [($M/G/1$): (inf/GD model)] queuing systems. Int. J. Grid Comput. Appl. **4** (2013)
4. K. Jayapriya, et al., An extensive survey on QoS in cloud computing. Int. J. Comput. Sci. Inf. Technol. **5** (2014)
5. S. Molinaro et al., Comparing expected wait times of an $M/M/1$ queue. Department of Mathematics and Statistics, University of Winsor (2010)
6. A. Tripathi, Simulation of queuing models. Int. J. Eng. Sci. Innov. Technol. **2** (2013)
7. R. Ravanmehr, et al., Cloud computing performance evaluation: issues and challenges. Int. J. Cloud Comput. Serv. Archit. **3** (2013)
8. T.V. Mathew, Queuing analysis, transportation systems engineering. Indian Institute of Technology, Bombay, Feb 2014. D. Praveen, K. Satish, The queuing theory in cloud computing to reduce the waiting time. IJCSET **1** (2011)
9. H. Wang, et al., A conditional probability approach to $M/G/1$-like queues. Perform. Eval. **65** (2008)
10. W. Stallings, Queuing analysis (2000)
11. H. Groenvelt, A note on queuing models (1996)
12. J. Sztrik, et al., Basic queuing theory. University of Debrecen, Faculty of Informatics (2012)

Soft Computing and Big Data Intelligence for a Low-Carbon Economy

Jason Levy

1 Introduction

As global climate risks are increasing, there is a need for big data intelligence and soft computing to deal with the challenging and complex issues associated with promoting a low-carbon future. A large body of scholarship is devoted to managing the complexity of large-scale, complex, data rich neuro-fuzzy systems. For example, Verleysen and Damien [19] and Brown et al. [4] address the 'curse of dimensionality' in high-dimensional neuro-fuzzy systems. Neuro-fuzzy inference systems constitute a valuable tool for modeling the transition to a decarbonized economy as they constitute a combination of two basic concepts: fuzzy inference systems (a non-adaptive system that is built using a collection of fuzzy rules that may have a semantic interpretation) and neural network systems (highly adaptive networks of components that generate superficial models, but lack in semantic interpretability). While combing these two systems can lead to greater efficiency [10], this has generally occurred in relatively straightforward applications that are primarily low order [8, 10]. However, there is now great urgency to deal with complex sustainability initiatives, such as the transition to a low-carbon economy. These challenges demand greater efficiency as they constitute complex, higher-order, time-critical applications. Accordingly, to promote a decarbonized economy in Hawaii and—ultimately the world—we develop a new tensor field neuro-fuzzy model. This new model retains the essential dual characteristics of the fuzzy-neural models, which are now referred to as vector field neuro-fuzzy models (to distinguish them from our proposed new approach). In traditional fuzzy-neural models, an overall function of the individual vector field strengths produces the classifications, inferences, and universal function approximations. Given an input vector, the traditional neuro-fuzzy approach proceeds in sequential order until every component is evaluated. An exhaustive evaluation of rules, radial basis functions,

---

J. Levy (✉)
University of Hawaii, Honolulu, HI, USA
e-mail: jlevy@hawaii.edu

or activation functions is carried out, often followed by a 'defuzzification' process. For example, Chiu [5] extracts fuzzy classification rules by subtractive clustering. For enhanced efficiency, these tasks can be carried out in parallel on multiprocessing machines [2, 4], by dividing the vector filed function evaluations among processors. Other proposed methods of dealing with combinatorial rule explosion and complexity in fuzzy logic engines involve using union rule configuration (URC) instead of intersection rule configuration (IRC) [6, 7, 18]

The traditional neuro-fuzzy model is essentially a set of vector functions **F** each with sets of non-translational adaptive parameters **P** (i.e., those parameters that do not translate or move the entire function in space, for example, 'left or right' in the case of a 2D function, but change its shape) and translational adaptive parameters **Q**, on universes of discourse **U**, and a set of interconnections **I** (i.e., parameters of an output function) that produce the scalar network output function f. In other words, all vector field models can be stated succinctly as follows:

$$\forall \mathbf{f} \in \mathbf{F} \cdot \forall \mathbf{Q} \cdot \exists (\mathbf{p} \in \mathbf{P} \wedge \mathbf{x} \in \mathbf{U}) \cdot (f(\mathbf{x}, \mathbf{P}, \mathbf{Q}, \mathbf{I}) \neq 0) \quad (1)$$

That some parameters exist for any given input that makes the function nonzero defines its continuity. This is true of a network of any of the following functions:

$$\text{Generalized bell } f(x, \{a, b, c\}) = 1/(1 + |(x - c)/a|^{2b}), \quad P = \{a, b\} \quad (2)$$

$$\text{Gaussian } f(x, \{c, \sigma\}) = \exp(-0.5[(x - c)/\sigma]^2), \quad P = \{\sigma\} \quad (3)$$

$$\text{Triangular } f(x, \{a, b, c\}) = \max(\min((x - a)/(b - a), (c - x)/c - b)), 0),$$
$$P = \{a, b, c\} \quad (4)$$

$$\text{Trapezoidal } f(x, \{a, b, c, d\}) = \max(\min((x - a)/(b - a), 1, (d - x)/d - c)), 0),$$
$$P = \{a, b, c, d\} \quad (5)$$

$$\text{Sigmoidal } f(x, \{a, c\}) = 1/(1 + \exp(-a(x - c))), \quad P = \{a\} \quad (6)$$

$$\text{Signum } f(\mathbf{x}, \{\mathbf{w}, \theta\}) = \begin{cases} 1 & \mathbf{wx} + \theta > 0 \\ -1 & \text{otherwise} \end{cases}, \quad P = \{\theta\} \quad (7)$$

$$\text{Logistic } f(\mathbf{x}, \{\mathbf{w}, \theta\}) = 1/(1 + \exp(-\mathbf{wx} - \theta)), \quad P = \{\theta\} \quad (8)$$

$$\text{Hyperbolic Tangent } f(\mathbf{x}, \{\mathbf{w}, \theta\}) = (1 - \exp(-\mathbf{wx} - \theta))/(1 + \exp(-\mathbf{wx} - \theta)),$$
$$P = \{\theta\} \quad (9)$$

$$\text{Radial Basis Function } f(\mathbf{x}, \{\mathbf{u}, \sigma\}) = R(||\mathbf{x} - \mathbf{u}||/\sigma) \quad P = \{\sigma\} \quad (10)$$

$$\text{Step } f(\mathbf{x}, \{w, \theta\}) = \begin{cases} 1 & \mathbf{wx} + \theta > 0 \\ 0 & \text{otherwise} \end{cases}, \quad P = \{\theta\} \quad (11)$$

Systems that use any of the functions in the above list, or other functions that satisfy the condition of Eq. (1), can be broadly classified as continuous systems in the literature. These constitute majority of current neuro-fuzzy systems. Take the typical generalized bell function (Eq. 2) used in neuro-fuzzy models for example (Fig. 1).

Given translational parameter $c = 50$, it is possible to make this function nonzero for any input by changing adaptable non-translational parameters a and b. Similarly for the Gaussian type functions in the list (which includes the radial basis function), given a specific translational mean as a parameter changing the adaptive variance parameter (σ) can make the function nonzero for any input. The same applies to the step and logistic functions where for given weights ($\mathbf{W}$) (which positions the function in space), a threshold (θ) can also be found to make these functions nonzero. The current literature emphasizes the continuous nature of neuro-fuzzy models. Many of the above functions in Eqs. 2–11 can be made discrete by setting the parameter that determines its span (or width) to be fixed instead of adaptive. While this modification is relatively straightforward, the only examples found in the literature are those using the step and signum functions (Eqs. 7 and 11) where by setting fixed thresholds, functions become discrete [1, 3, 11–13]. In addition, Aleksander and Morton [1] and Austin [3] discuss the implementation of the aforementioned discrete functions implemented as random-access memory (RAM)

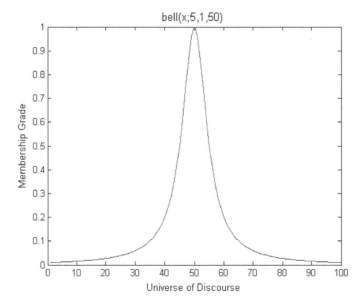

Fig. 1 Typical generalized bell membership function

1.1 Tensor Field Neuro-Fuzzy Model

A tensor field is a variable geometric quantity in m-dimensional space represented by an array with n-indices indicating its rank. The simplest geometric quantity is a scalar which is a tensor of rank zero (no indices) followed by a vector which is a tensor of rank one (one index) then a matrix which has two indices (tensor of rank two) followed by higher-ranked tensors. Apart from the number of indices, tensors differ in how they transform for a change of coordinate system (under rotation, for example). It is known that vectors transform differently than scalars and matrices transform differently than vectors. The neuro-fuzzy tensor constitutes a rank-two (i.e., matrix) tensor where each component of the tensor is a vector. It is proposed that this formulation of neuro-fuzzy tensor field problems (rather than using vector field models) offers important advantages in computational efficiencies as resolving vectors into orthogonal components makes the computation more efficient. For example, if a vector is orthogonal (perpendicular) to axes of its coordinate system, then it is known that there is no need to compute these components, which would be zero.

Figure 2 illustrates a neuro-fuzzy system comprised of two rules with two inputs using both a vector field neuro-fuzzy model (Fig. 2a) and a tensor field neuro-fuzzy model (Fig. 2b). As inputs to the vector field neuro-fuzzy network change (x_1, x_2) (following the curves shown in Fig. 2), rules fire with varying strengths. These strengths are indicated by the length of the vectors which point toward the location of the rule (Fig. 2a). The same applies for the tensor field model except that both vectors can be incorporated into the same axes (shown as little triangles). Similar information is graphically conveyed in both models. Products of generalized bell membership functions (Eq. 2), one for each axis (centered on rules typical of neuro-fuzzy computation), were used to generate information about the vectors in Fig. 2.

2 Tensor Field Neuro-Fuzzy Models and the Modeling a Low-Carbon Economy

The tensor field neuro-fuzzy model defines a vector of vectors (i.e., a matrix tensor field). Components of the matrix tensor are vectors, while components of the vectors are scalars. The orthogonality of a sample tensor **X** and another tensor **Y** is shown in Eq. 14. Note that dot products of orthogonal vectors and tensors are zero $\mathbf{X} \cdot \mathbf{Y} = x_i y_i$ (Einstein summation notation). The analysis which follows constitutes an original extension and reformulation of concepts in previously published work [14–16] applied to the chaotic Mackey Glass process [9, 17] as a benchmark.

The dimensions m of one index $i = 1, \ldots, m$ of the neuro-fuzzy matrix tensor (its number of vector components) is equal to the number of rules and that of the other index $j = 1, \ldots, n$ (the number of scalar components of each vector) is equal to the number of network inputs n. If dimensions n and m are unbounded, and then when n

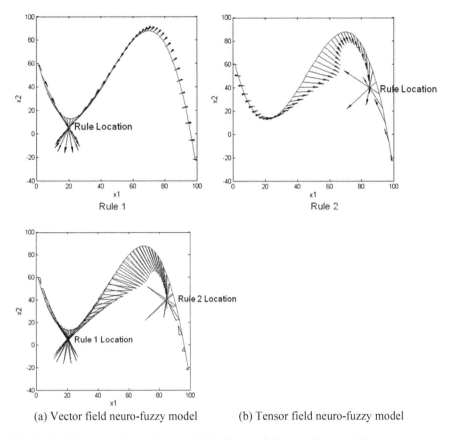

(a) Vector field neuro-fuzzy model   (b) Tensor field neuro-fuzzy model

Fig. 2 Graphical comparisons of vector field and tensor field neuro-fuzzy models

is very large, the product part of the product-sum neuro-fuzzy composition (Eq. 12) over membership functions not near unity for a given **x** produces component vectors so small that they can effectively be approximated to zero (i.e., $w_i < \varepsilon$, ε is some very small number). In their initial states, bell or Gaussian-shaped membership functions have near unity values in a very narrow range (see Fig. 1, for example). In practice, this means that if there are a large number of components (m) with **c** distributed about the space, then most components will be zero with only a few nonzero components, i.e., those with membership functions very close to unity. The result is a field $\mathbf{T}(\mathbf{x})$ composed of sparse tensors, the Kronecker product of a sparse weight vector and unity tensor. As the tensor field strength changes as inputs **x** change, zero vector components would switch to nonzero as rules are approached, and nonzero vector components switch to zero as other rules simultaneously recede. The zero value for tensor dot product of changing tensors indicates orthogonality (14). The resulting field can be described as a sparse orthogonal tensor field.

$$w_i(\mathbf{x}, \{\mathbf{a}, \mathbf{b}, \mathbf{c}\}) = \prod_{j=1}^{n} 1/(1 + |(x_j - c_{ij})/a_{ij}|^{2b_{ij}}), \quad i = 1, \ldots, m \quad (12)$$

$$\mathbf{T}(\mathbf{x}) = \mathbf{w} \otimes \mathbf{r} = \begin{bmatrix} w_1 \\ w_2 \\ \vdots \\ w_m \end{bmatrix} \otimes \begin{bmatrix} (\mathbf{x} - \mathbf{c}_1)/\|\mathbf{x} - \mathbf{c}_1\| \\ (\mathbf{x} - \mathbf{c}_2)/\|\mathbf{x} - \mathbf{c}_2\| \\ \vdots \\ (\mathbf{x} - \mathbf{c}_m)/\|\mathbf{x} - \mathbf{c}_m\| \end{bmatrix} = \begin{bmatrix} w_1(\mathbf{x} - \mathbf{c}_1)/\|\mathbf{x} - \mathbf{c}_1\| \\ w_2(\mathbf{x} - \mathbf{c}_2)/\|\mathbf{x} - \mathbf{c}_2\| \\ \vdots \\ w_m(\mathbf{x} - \mathbf{c}_m)/\|\mathbf{x} - \mathbf{c}_m\| \end{bmatrix}$$
$$(13)$$

$$\begin{bmatrix} w_1(\mathbf{x} - \mathbf{c}_1)/\|\mathbf{x} - \mathbf{c}_1\| \\ 0 \\ \vdots \\ 0 \end{bmatrix} \cdot \begin{bmatrix} 0 \\ w_2(\mathbf{x} - \mathbf{c}_2)/\|\mathbf{x} - \mathbf{c}_1\| \\ \vdots \\ 0 \end{bmatrix} = 0 \quad (14)$$

$$y = \mathbf{w} \cdot \mathbf{f} = \sum_{i=1}^{m} w_i f_i \approx \sum_{\{i \mid w_i > \varepsilon\}} w_i f_i \quad (15)$$

2.1 Exploiting Sparsity and Orthogonality in Tensor Field Models

A sparse orthogonal tensor field results in advances in computational efficiency. Specifically, once the tensor indices of zero components are known, then these components do not have to be computed in producing the neuro-fuzzy scalar output y (shown in Eq. 15). The orthogonality property ensures that there will always be the same number of zero components, regardless of how tensors change throughout the field. This constitutes a major difference with conventional neuro-fuzzy systems. Conventional soft computing systems do not arrange vector field components into an ordered tensor and hence cannot exploit properties of sparsity and orthogonality of the resulting tensors. Accordingly, all traditional fuzzy-neural systems must evaluate vector components exhaustively and usually in a sequence (on von Neumann architecture machines). Lacking tensor indices, traditional soft computing systems are unaware of which components need not be computed (i.e., the zero components). Indeed, in traditional soft computing, one often discovers after the fact (of exhaustively computing vector field components) that only a few rules file (i.e., there are only a small number of nonzero components).

This exhaustive computation is a major problem only if there are very many computations to perform, i.e., very many rules, or complexity that requires very many vector field components and may go unnoticed. However, identifying optimal scenarios for a sustainable energy future in Hawaii and promoting a global low-carbon economy are problems characterized by high complexity and high order.

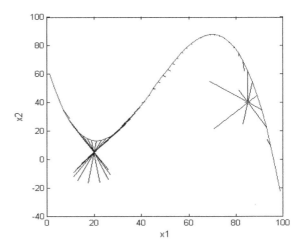

Fig. 3 Simulated high-order neuro-fuzzy tensor field showing orthogonal tensors

Accordingly, the increased efficiency of the new tensor field model ($|\{i|w_i > \varepsilon\}| \ll m$ in the computation of Eq. 15) is critical. Note however that Eq. 15 applies to all systems of any order or complexity. However, computing the nonzero components would only be absolutely necessary in the high complexity, high-order problems previously discussed. Figure 3 illustrates the orthogonality effect (with the tensors field component vectors computed for $n = 50$) calculated using generalized bell membership function (shown in Fig. 1 to the power 50 instead of 2).

2.2 Tensor Field Models: Sparsity, Orthogonality, and Discrete Vector States

In the previous section, the conditions of sparsity and orthogonality were discussed. Recall that sparsity and orthogonality hold when the initial condition of membership functions is unity or close to unity within a bounded core. For example, consider the generalized bell membership function (Eq. 2) with adaptive parameters P that determine the width of the core. Parameter a of P is primarily responsible for the width of the bell, which is the distance measured between the two crossover membership values of 0.5 of the function, which is a constant value for fixed a. The slope at the crossover point is determined by b. To limit the width of this core, it is necessary to have a fixed parameter, denoted by a, while parameters b and c can remain adaptive. This represents a paradigm shift from conventional neuro-fuzzy computations and is necessary to achieve a bounded core and produce the sparsity and orthogonal conditions of the neuro-fuzzy tensor field model.

Given a high-order system and a fixed with **a,** the resulting output multidimensional membership function (produced by the product of many generalized bell shaped functions) resembles a pulse. The model can hence be described as discrete (rather than continuous) since the model is broken up into individually distinct orthogonal components, with each component defining the model locally. The corollary of Eq. (1) for discrete systems is given by:

$$\neg \forall \mathbf{f} \in \mathbf{F} \cdot \forall \mathbf{Q} \cdot \exists (\mathbf{p} \in \mathbf{P} \wedge \mathbf{x} \in \mathbf{U}) \cdot (f(\mathbf{x}, \mathbf{P}, \mathbf{Q}, \mathbf{I}) \neq 0) \tag{16}$$

For some input **x,** no selection of parameters **P** exists that could produce a nonzero scalar output function. Establishing a nonzero output, one must adjust **Q**, thereby translating components about the space. These properties naturally manifest themselves in high-order tensor fields. A simple modification to the generalized bell membership function used in conventional neuro-fuzzy can result in enhanced orthogonality for both low- and high-order problems and be used to graphically demonstrate the low-order pulse formed ($\beta \gg 1$). This modification to the generalized bell membership function is given as

$$f(x, \{a, b, c\}) = 1/(1 + \beta|(x - c)/a|^{2b}), \quad P = \{b\} \tag{17}$$

Continuous and discrete generalized bell functions are provided in Fig. 4. A continuous bell function with $\beta = 1$ is provided in Fig. 4a, whereas a discrete bell function with $\beta = 100$ is provided in Fig. 4b. In both figures, the parameters **P** change (see Eq. 13 for more details). Note that Fig. 4a is not restricted to a narrow pulse as **P** changes, whereas Fig. 4b is effectively always zero outside of the region within the pulse.

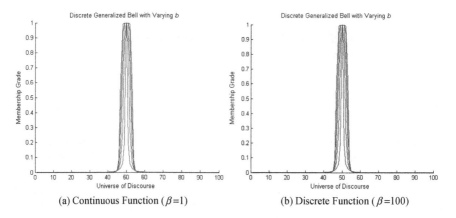

(a) Continuous Function ($\beta=1$) (b) Discrete Function ($\beta=100$)

Fig. 4 Continuous and discrete generalized bell functions

3 Tensor Field Model Implementation Using Sparse Multidimensional Arrays

A neuro-fuzzy model can be described as a rank-two tensor field (matrix tensor field) of dimensions $m \times n$. The space is thus n-dimensional, and tensors have m vector components that vary at different points $\mathbf{x}$ in the space. This space is modeled by using a sparse rectangular n-dimensional array $A_{i_1 \times i_2 \times \cdots \times i_n}$. A full array would be difficult to implement without exceeding virtual memory limits, so a very sparse array assists with the computations. The sizes of the dimensions i_j, $j = 1, \ldots, n$ will affect the precision of modeling efforts.

The nonzero components of the array mark the location points of rule centers $\mathbf{c}_i$, $i = 1, \ldots, m$ and the reference points for calculating tensor field vector components $w_i(\mathbf{x} - \mathbf{c}_i)/\|\mathbf{x} - \mathbf{c}_i\|$, which is the vector component weight w_i multiplied by the unit vector $(\mathbf{x} - \mathbf{c}_i)/\|\mathbf{x} - \mathbf{c}_i\|$ pointing toward the center of the rule. The nonzero components of the tensor $w_i \neq 0$ can be found for a given $\mathbf{x}$ by looking up the array at $\mathbf{x} \pm \mathbf{a}_r$, $\mathbf{a}_r < \mathbf{a}$ for nonzero components. This is a sub-hypercube of the array of side $2a_r$ centered on $\mathbf{x}$. At addresses $\mathbf{c}_i$, $i = 1, \ldots, m$ of the array are stored parameters $\mathbf{b}_i$, $i = 1, \ldots, m$ further defining the tensor field and $\mathbf{p}_i$, $i = 1, \ldots, m$ of the scalar field. A multidimensional array is also referred to as tensor, and our research involves both n-dimensional sparse array tensor implementations and the two-dimensional tensor of the field.

3.1 Neuro-Fuzzy Tensor Field Calculus

Neuro-fuzzy networks produce scalar outputs y (Eq. 15) for given input vector $\mathbf{x}$. The goal of training a neuro-fuzzy network is to produce a scalar field that optimally fits both the training data and checking data. This can be termed a superficial (surface or manifold) model. However, the scalar field is affected by properties of the tensor field (Eqs. 13 and 15). To optimize the scalar field, it may be necessary to adjust the tensor field (Eq. 13). The scalar field therefore has an underlying tensor field model. This model can be termed a semantic model, because it can be used to derive a language that describes the superficial model made up of terms given to various configurations of the tensor field. The configuration of the tensor field after training can reveal the semantics of the model including the number of rules discovered (if any). The algorithm that optimizes the scalar field by adjusting the tensor field is described in the Appendix and in Fig. 5. This algorithm describes how parameters $[\mathbf{c}_i, \mathbf{b}_i, \mathbf{p}_i]$, $i = 1, \ldots, m$ are adjusted in minimizing a quadratic error function. Parameter vectors $\mathbf{c}$ give the locations of nonzero components in the sparse array. As these vectors are updated per epoch, their new values give the new locations of the components in the array and the array is modified by moving these components to the new locations. Changes to $\mathbf{b}$ affect the magnitude of vector components of tensors in the field. Changes to $\mathbf{p}$ only affect the scalar field. As components move

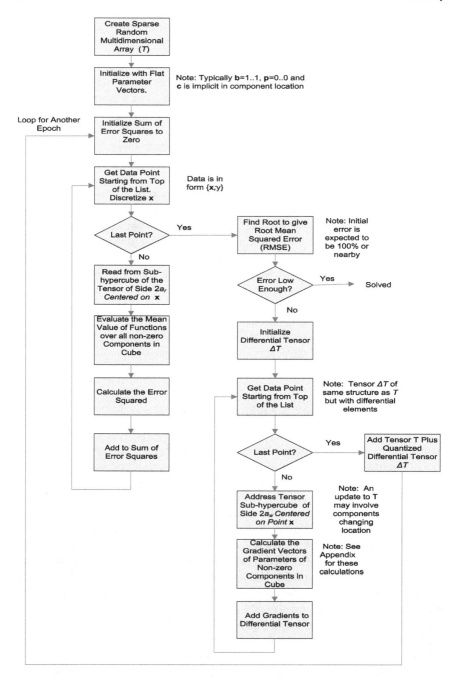

Fig. 5 Neuro-fuzzy tensor calculus

about the array during training, it is proposed that clustering of components can be interpreted as rule formation.

Tensor T referred to in Fig. 5 is a sparse multidimensional array, where the locations of nonzero elements in the array **c** are adaptive parameters, which are sparse. They can be given initial random scatter locations in the sparse array or some special configuration, if there is some a priori semantic knowledge. These locations store parameters **b** and **p** which are also adaptive. Training proceeds by processing all data points iteratively, accumulating differential changes to the **c**, **b**, and **p** parameters calculated in a differential tensor ΔT until the end of the epoch. The next step involves adding those differences to the original tensor ($T^i = T^i + \Delta T^i$) so that the updated model contains less error. This process is repeated for each epoch i, until a sufficiently low error is achieved.

4 Comparing Tensor Field and Vector Field Neuro-Fuzzy Model Performance for Promoting a Low-Carbon Economy

A comparison of performance is given for MATLAB's classic fuzzy logic toolbox function ANFIS [10] and is referred to as System 1 in Table 1 (vector field fuzzy-neural model). System 1 is compared with the proposed tensor neuro-fuzzy model (denoted by System 2) which was also encoded using MATLAB. There are two major options in MATLAB's ANFIS toolbox: genfis1 and genfis2. The number of rules assumed in genfis1 is determined by the partitioning of inputs into a specific number of fuzzy sets. The number of rules assumed in genfis2 is determined by the selection of radii of influence of data cluster centers. Another difference is that the number of rules in genfis1 cannot be arbitrarily reduced to a small number, as is the case of genfis2.

One of the problems with applying traditional soft computing methods to the modeling of a low-carbon economy is that many rules are likely discovered as opposed to being a priori selected. Knowledge discovery can range from an arbitrary large number of rules (which can be as large as the number of data points used) to a minimum of one rule. In fact, most real-world energy and economic modeling problems contain unknown systems or processes that involve significant knowledge and rule discovery (rather than simply parameter estimation of a pre-determined model). An advantage of the tensor field neuro-fuzzy model is that no rules are assumed a priori, but rather rules are discovered after training by the clustering of components in the tensor (sparse n-dimensional array).

Both systems were trained with exactly the same data for a number of epochs on a Dell Optiplex GX745 Intel® Core™ 2 CPU 6300 machine at 1.86 GHz with 1.99 GB of RAM running a Microsoft Windows XP Professional Version 2002 Service Pack 2 and MATLAB Version 7.0.1.24704 (R14) Service Pack 1. The training time and recall time for both systems are compared for various order problems. The machine

Table 1 Comparison of ANFIS and tensor field neuro-fuzzy training and evaluation

RMS output	RMSE		Training time (s)		Evaluation time (s)		No of inputs	No of data points	No of epochs
	System 2 tensor	System 1 vector	System 2 tensor	System 1 vector	System 2 tensor	System 1 vector			
97.09	14.7052	8.39	23.11	0.031	0.002	0.0008	2	40	50
87.46	11.15	11.96	80.5	0.219	0.0043	0.000775	3	80	50
75.64	15.51	13.9582	343.76	2.141	0.0099	0.000681	4	160	50
85.16	8.75	14.2074	1846.97	23.00	0.0259	0.000781	5	320	50
94.31	9.18	0.0074	2230.13	142.51	0.0323	0.187	6	320	50
79.05	14.9854	0.00018	2658.26	834.297	0.0378	0.0011	7	320	50
83.96	11.0829	0.00004	3065.92	4269.156	0.0428	0.016	8	320	50
106.22	13.84	0.00003	4743.98	21,719.39	0.068	0.016	9	320	50
92.56	13.8	0.35	4972.73	10,954.34[a]	0.088	0.105	10	320	50
103.65	11.24	6.523	6732.01	20,351.78	0.092	0.125	11	320	50
98.34	9.12	No data	12317.98	No data[b]	0.100	No data	15	640	50
75.80	11.08	No data	17719.16	No data	0.144	No data	20	640	50
76.963	8.29	No data	38471.58	No data	0.2744	No data	30	1280	50

[a]Switch from genfis1 to genfis2. *RMS*-root mean square value
[b]Solution time too long. *RMSE*-root mean squared error

used for testing System 1 produces an 'out of memory' error and stops for tenth-order problems and higher using genfis1. This means that a solution with the stated computational system cannot be obtained for problems of tenth order and higher using ANFIS genfis1. Accordingly, for problems of tenth order and beyond, only genfis2 could be used. Switching to genfis2 however meant choosing a variable called 'the radius of influence of cluster centers.' The solution time and training errors varied widely with the chosen radius. The results are shown in Table 1. In higher-order problems as the size of the space grows exponentially, the number of rules is likely growing exponentially as well. Accordingly, for these higher-order problems, obtaining data becomes impractical in system 1: The growing inefficiencies contribute to excessive computational times that are indicated by 'no data' in Table 1.

5 Conclusions

The results obtained from a comparative study of performance during training and evaluation, with a wide range of green energy and environmental datasets, suggest that the new tensor field fuzzy-neural model is more efficient than the conventional vector field model, thereby allowing an analyst to better understand and predict highly complex data and real-world problems at the interface of the environment and society. Moreover, the new tensor field model involves exploiting the sparsity and orthogonality of tensors in a field which is not possible with the independent vector field models of conventional neuro-fuzzy inference systems. As well, the more comprehensive analysis and semantic modeling of the underlying neuro-fuzzy tensor field indicate a greater potential for more accurate knowledge discovery and then pertains to the conventional model.

Finally, while the tensor model remains functional as the order of the economic problems get higher, the errors are large in comparison with the conventional fuzzy-neural model (system 1 in Table 1). This is due to ill-conditioning of the problem as result of a lack of normalization of weights. That is, w_i are used in Eq. (15) rather than normalized weights $w_i/(w_1 + w_2 + \cdots + w_m)$ of the conventional model. If normalized weights are used, errors are generally reduced to levels as low as that of classic neuro-fuzzy systems (system 1). A proposed solution is to begin with absolute weights (which positions clusters in the right locations) and then switch to normalized weights to address the ill-conditioning (which should produce better convergence and lower errors).

The implications of these soft computing models are profound for guiding policy makers toward a very low-carbon economy in Hawaii—and around the world. The future of sustainability in Hawaii rests in hands of business leaders, decision makers, government officials, and local communities that are willing to engage in the planning process, promote a green economy, demand regulatory action, and embrace stewardship. Specifically, understanding the preferences and values of all affected parties interested in promoting a low-carbon economy can help economists, decision makers, and climate change professionals to understand the key administrative, management, and public policy issues. However, in traditional soft computing models, incorporating all data sources needed to address the problem of decarbonizing Hawaii's economy leads to the 'curse of dimensionality' (excessive computational times, growing inefficiencies, etc.).

Appendix

Neuro-fuzzy Tensor Calculus

$$e(\theta) = \sum_{k=1}^{M} (y(\theta)_k - \hat{y}_k)^2, \quad \theta = [\theta_c, \theta_b, \theta_p] = [\mathbf{c}, \mathbf{b}, \mathbf{p}] \ (M \text{ data points}) \quad (18)$$

Goal is to find θ^*

$$\theta^* = \arg \min e(\theta) \text{ (i.e., the parameters resulting in minimum error)} \quad (19)$$

$$e_k(\theta) = (y(\theta)_k - \hat{y}_k)^2 \text{ (Error squared for a single point)} \quad (20)$$

$$y(\theta)_k = \sum_{i=1}^{m} w_i f_i$$
$$= w_1[p_{11}(d(x_1) - c_{11}) + p_{12}(d(x_2) - c_{12}) + \cdots + p_{1n}(d(x_n) - c_{1n}) + p_{1(n+1)}]$$
$$+ w_2[p_{21}(d(x_1) - c_{21}) + p_{22}(d(x_2) - c_{22}) + \cdots + p_{2n}(d(x_n) - c_{2n}) + p_{2.n+1}] + \cdots$$
$$+ w_m[p_{m1}(d(x_1) - c_{m1}) + p_{m2}(d(x_2) - c_{m2}) + \cdots + p_{mn}(d(x_n) - c_{mn}) + p_{m(n+1)}] \quad (21)$$

$$w_i = \prod_{j=1}^{n} w_{ij} = \prod_{j=1}^{n} \frac{1}{1 + \beta \left| \frac{d(x_j) - c_{ij}}{a_r} \right|^{2b_{ij}}} \quad (22)$$

$$\frac{\partial e_k}{\partial c_{ij}} = 2(y(\theta)_k - \hat{y}_k) \left[f_i \frac{\partial w_i}{\partial c_{ij}} + w_i \frac{\partial f_i}{\partial c_{ij}} \right] \quad (23)$$

$$f_i = p_{i1}(d(x_1) - c_{i1}) + p_{i2}(d(x_2) - c_{i2}) + \cdots + p_{in}(d(x_n) - c_{in}) + p_{i(n+1)} \quad (24)$$

$$\frac{\partial f_i}{\partial c_{ij}} = -p_{ij}, \quad \frac{\partial w_i}{\partial c_{ij}} = \begin{cases} \frac{2b_{ij}}{d(x_j) - c_{ij}} w_{ij}(1 - w_{ij}) & x_j \neq c_{ij} \\ 0 & x_j = c_{ij} \end{cases} \quad (25)$$

$$\frac{\partial e_k}{\partial b_{ij}} = 2(y(\theta)_k - \hat{y}_k) f_i \frac{\partial w_i}{\partial b_{ij}} \quad (26)$$

$$\frac{\partial w_i}{\partial b_{ij}} = \begin{cases} -2\ln \left| \frac{d(x_j) - c_{ij}}{a_r} \right| w_{ij}(1 - w_{ij}) & x_j \neq c_{ij} \\ 0 & x_j = c_{ij} \end{cases} \quad (27)$$

$$\frac{\partial e_k}{\partial p_{ij}} = 2(y(\boldsymbol{\theta})_k - \hat{y}_k) w_i \frac{\partial f_i}{\partial p_{ij}}, \quad \frac{\partial f_i}{\partial p_{ij}} = \begin{cases} (d(x_j) - c_{ij}) & j \leq n \\ 1 & j = n+1 \end{cases} \tag{28}$$

$$\nabla e_k(\boldsymbol{\theta}) \equiv \left[\frac{\partial e}{\partial c_{11}}, \frac{\partial e}{\partial c_{12}}, \ldots, \frac{\partial e}{\partial c_{1n}}, \frac{\partial e}{\partial c_{21}}, \frac{\partial e}{\partial c_{22}}, \ldots, \frac{\partial e}{\partial c_{2n}}, \ldots, \frac{\partial e}{\partial c_{m1}}, \frac{\partial e}{\partial c_{m2}}, \ldots, \frac{\partial e}{\partial c_{mn}}, \right.$$

$$\frac{\partial e}{\partial b_{11}}, \frac{\partial e}{\partial b_{12}}, \ldots, \frac{\partial e}{\partial b_{1n}}, \frac{\partial e}{\partial b_{21}}, \frac{\partial e}{\partial b_{22}}, \ldots, \frac{\partial e}{\partial b_{2n}}, \ldots, \frac{\partial e}{\partial b_{m1}}, \frac{\partial e}{\partial b_{m2}}, \ldots, \frac{\partial e}{\partial b_{mn}},$$

$$\left. \frac{\partial e}{\partial p_{11}}, \frac{\partial e}{\partial p_{12}}, \ldots, \frac{\partial e}{\partial p_{1,n+1}}, \frac{\partial e}{\partial p_{21}}, \frac{\partial e}{\partial p_{22}}, \ldots, \frac{\partial e}{\partial p_{2,n+1}}, \ldots, \frac{\partial e}{\partial p_{m1}}, \frac{\partial e}{\partial p_{m2}}, \ldots, \frac{\partial e}{\partial p_{m,n+1}} \right]^T \tag{29}$$

$$\nabla e(\boldsymbol{\theta}^i) = \sum_{k=1}^{M} \nabla e_k(\boldsymbol{\theta}^i) \quad \text{(Gradient over all } M \text{ points–batch learning)} \tag{30}$$

$$\boldsymbol{\theta}^{i+1} = \boldsymbol{\theta}^i - f(\nabla e(\boldsymbol{\theta}^i)), \quad f(\nabla e_k(\boldsymbol{\theta}^i))^T = [\text{sgn}(\nabla e(\boldsymbol{\theta}^i)_{\theta=c}) \cdot q_c \, \eta \nabla e_k(\boldsymbol{\theta}^i)_{\theta \neq c}] \tag{31}$$

where
$q_c = d^{-1}(i+1) - d^{-1}(i)$ and η is a learning rate for parameters not quantized and $d : \mathbf{x} \to \mathbf{i}, \mathbf{i} = [i_1, i_2, \ldots, i_n], i_i \in [1, 2, \ldots, N]$ are tensor indices.

These calculations are looped for several epochs $i = 1, \ldots, N$ giving $\boldsymbol{\theta}^N$

$$\boldsymbol{\theta}^N \to \boldsymbol{\theta}^*, \quad N \to \infty \tag{32}$$

References

1. I. Aleksander, H. Morton, *Neurons and Symbols—The Stuff that Mind is Made of* (Chapman Hall, London, 1993)
2. J. Andrews, *Taming Complexity in Large-Scale Fuzzy Systems* (PC AI, Phoenix: Knowledge Technology Inc., 1997), pp. 39–42
3. J. Austin (ed.), *Ram-Based Neural Networks. Progress in Neural Processing*, vol. 9 (World Scientific, 1998), pp. 18–30
4. M. Brown, K.M. Bossley, D.J. Mills, C.J. Harris,(1995) High dimensional neurofuzzy systems: overcoming the curse of dimensionality, in *Proceedings of 1995 IEEE International Conference on Fuzzy Systems, 1995. International Joint Conference of the Fourth IEEE International Conference on Fuzzy Systems and The Second International Fuzzy Engineering Symposium*, vol. 4. Yokohama, Japan, 20–24 Mar 1995, pp. 2139–2146
5. S., Chiu, Method and software for extracting fuzzy classification rules by subtractive clustering, in *Biennial Conference of the North American Fuzzy Information Processing Society—NAFIPS* (Cat. No. 96TH8171), Berkeley, CA, USA, 461-5. NAFIPS—North American Fuzzy Information Processing Society. BISC—Berkeley Initiative in Soft Computing IEEE Neural Networks Council. IEEE Systems, Man and Cybernetics Society, June 1996, pp. 19–22
6. W. Combs, Reconfiguring the fuzzy rule matrix for large, time-critical applications, in *Third Annual International Conference on Fuzzy-Neural Applications, Systems and Tools*, vol. 18 (PennWell Publishing Company, Nashua, 1995), pp. 1–7
7. W. Combs, J. Andrews, Combinatorial rule explosion eliminated by a fuzzy rule configuration. IEEE Trans. Fuzzy Syst. **06**(01), 1–11 (1998)

8. E. Cox, M. O'Hagen, R. Taber, *The Fuzzy Systems Handbook, A Practitioner's Guide to Building, Using, and Maintaining Fuzzy Systems* (Morgan Kaufmann, 1998)
9. J. Gleik, *Chaos—Making a New Science* (Heinemann, London, 1988)
10. J.-S.R. Jang, C.-T. Sun, E. Mizutani, *Neuro-Fuzzy and Soft Computing—A Computational Approach to Learning and Machine Intelligence* (Prentice Hall, Upper Saddle River, NJ, 1997)
11. P. Kanerva, The spatter code for encoding concepts at many levels, in *ICANN'94, Proceedings of the International Conference on Artificial Neural Networks* (*Sorrento, Italy*) (Springer, London, 1994)
12. P. Kanerva, *Sparse Distributed Memory* (The MIT Press, Cambridge, MA, 1988)
13. P. Kanerva et al., *Computing with Large Random Patterns. Foundations of Real-World Intelligence* (CSLI, Stanford, CA, 2001)
14. A. Kong, Real sparse distributed memory. West Indian J. Eng. **25**(1), 52–62 (2002)
15. A. Kong, Impulse activated sparse cell array network in non-linear autoregressive process modeling. Soft Comput. Fusion Found. Methodol. Appl. **9**(6), 421–429 (2005) (Springer)
16. A. Kong, Sparse distributed fuzzy inference systems. Soft Comput. Fusion Found. Methodol. Appl. **10**, 567–577 (2006). (Springer)
17. M. Mackey, L. Glass, *From Clocks to Chaos: The Rhythms of Life* (New Jersey, 1988)
18. J.J. Weinschenk, J. Weinschenk, W.E. Combs, R.J. Marks, II, Avoidance of rule explosion by mapping fuzzy systems to a union rule configuration, in *The 12th IEEE International Conference on Fuzzy Systems*, vol. 1, (2003), pp. 43–48
19. M. Verleysen, F. Damien, *The Curse of Dimensionality in Data Mining. International Work-Conference on Artificial Neural Networks, Lecture Notes in computer Science*, vol. 3512 (Springer, 2005), pp. 758–770

Lung Image Classification to Identify Abnormal Cells Using Radial Basis Kernel Function of SVM

Sajja Tulasi Krishna and Hemantha Kumar Kalluri

1 Introduction

Nowadays, to provide better health for humans, the medical laboratory experts depend on the electronic healthcare systems for diagnosing the diseases and take necessary actions for treatment based on recommendations. Lung cancer [1] causes the abnormal growth of the cells named as malignant, and healthy growth of cells are called benign cells. Early identification of lung cancer and treatment may increase the life span of the patients who are suffered with lung cancer. With the rapid growth of technology, the lung cancer detection is an essential domain in the research.

Alsallal et al. [2] proposed a classification technique to classify the lung images; the researchers got 72.73% accuracy with RBF kernel. Saraswathi et al. [3] proposed an optimal critical point selection algorithm (OCPS) used for segmentation, and the suspected area nodule was found by the bidirectional chain code (BDC) classified using SVM. The researchers got 74.12% accuracy.

Devarapalli et al. [4] presented a survey about lung cancer detection with various existing methods on the LIDC dataset. The earlier researchers' work is shown on very fewer samples. But, today's world is accumulated with a large sample of data; to classify correct data from huge data is a crucial task. To overcome this limitation, the proposed method is tested on large datasets.

The remaining manuscript is organized as follows: This section explains the introduction and some related works. Section 2 describes the background methods. Section 3 explains the proposed approach; then, Sect. 4 gives experimental results. Conclusions are made in Sect. 5.

S. T. Krishna · H. K. Kalluri (✉)
Vignan's Foundation for Science, Technology and Research Deemed to be University, Vadlamudi, Guntur, AP, India
e-mail: hemanth_mtech2003@yahoo.com

2 Background

2.1 Pre-processing

This step enhances the image quality, which converts the RGB image into a gray image and after that removes noise by using the median filter, and finally, segmentation is done by the FCM algorithm [5] to extract the discriminative features.

2.2 Fuzzy C-means Algorithm

This algorithm is used to segment [6] the desired portion of the image.

1. Select initial fuzzy partition centroids, for clusters $C_0 = k_1$, $C_1 = k_2$, $C_2 = k_3$, ..., $C_k = k_n$ which is assigning values to all W_{ij}
2. Calculate the distance between corresponding cluster centers

$$\text{dist}(x_i, c_j) \qquad (1)$$

3. Repeat Steps 4 and 5 until the minimum j value is achieved
4. Then, the fuzzy partition, i.e., w_{ij}, is updated until the centroid of the clusters does not change. Here, 'm' is the fuzziness index.

$$W_{ij} = \frac{1}{\sum_{k=1}^{C}\left[\frac{\|x_i - c_j\|}{\|x_i - c_k\|}\right]^{\frac{2}{m-1}}} \qquad (2)$$

5. The centroids of each cluster are calculated by using the fuzzy partition.

$$c_j = \frac{\sum_{i=1}^{N} w_{ij}^m \cdot x_i}{\sum_{i=1}^{N} w_{ij}^m} \qquad (3)$$

FCM also minimizes the sum of squared error (SSE) and is given by

$$\text{SSE} = \sum_{j=1}^{k} \sum_{i=1}^{n} w_{ij}^p \text{dist}(x_i, c_j)^2 \qquad (4)$$

where p is a parameter that determines the influence of weights.

Fig. 1 Computation of LBP code

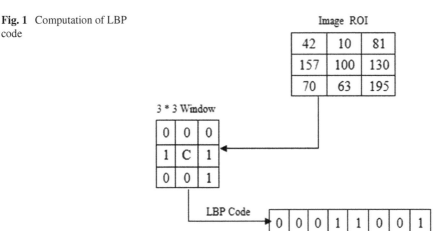

2.3 Local Binary Pattern (LBP)

LBP [7] divides an image into local pixels by thresholding the value of center pixels with each of the neighborhood pixels. The LBP feature is represented by using the gray values of eight neighborhood pixels. An LBP feature can be built by eight neighborhood pixels which are compared with the center pixel in one particular direction. This comparison gives either '1' or '0' [8]. A value '1' is assigned, when the pixel is ≥ to the center pixel, and a value '0' is assigned, when the pixel is < the center value. After this process, the decimal number of the corresponding binary pattern is replaced with the center pixel. The series of steps for calculation of LBP code is shown in Fig. 1.

2.4 Support Vector Machine

SVM classifies the data efficiently by using the hyperplane, but the hyperplane separates the linear data. In the real world, we can deal with the nonlinear data, so SVM used the kernel trick to separate the data.

The kernel function is defined as

$$K(\bar{x}) = \begin{matrix} 1 \text{ if } \|\bar{x}\| \leq 1 \\ 0 \text{ otherwise} \end{matrix} \qquad (5)$$

RBF function is localized, and it gives finite responses along with the x-axis. It has no prior knowledge about the data. It is defined as

$$k(x, y) = \exp\left(-\frac{\|x - y\|}{\sigma}\right) \qquad (6)$$

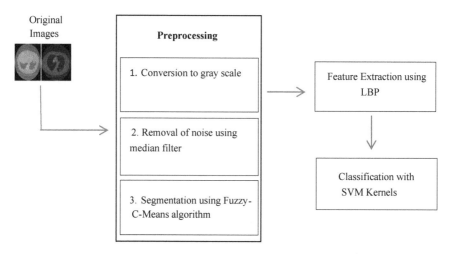

Fig. 2 Proposed architecture

3 Proposed Approach

To classify the usual and unusual tumors in the lung, various steps have been suggested and implemented by earlier researchers. The general methodology for classifying cancer usually includes pre-processing, feature extraction, and classification. In the pre-processing stage, convert the RGB images into the grayscale format. The median filter is applied to remove the noise. After for segmentation of the lung image, the fuzzy C-means algorithm is applied to extract the lung portion from the denoised lung images and discard the unnecessary portion from the image, i.e., remove the background information. In the feature extraction module, the Local Binary Pattern (LBP) algorithm is used to generate the feature vector from the segmented lung image. The extracted LBP features are used to train and to test, and the SVM classifier is used to classify whether they are benign or malignant images. The schematic diagram of the proposed work is depicted in Fig. 2.

4 Experimental Results and Discussion

4.1 Database

LIDC database [9] consists of computed tomography (CT) images of lungs. All the images of the dataset are in the DICOM format, which is used as a set standard in the medicine. According to the requirement, convert the images into.jpeg (Joint Photographic Experts Group) format. After that, separate the dataset into two groups as benign and malignant by using the XML file marked by radiologist which provided

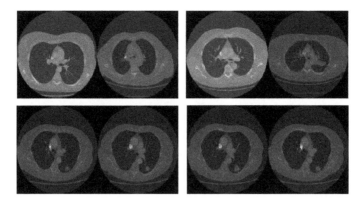

Fig. 3 Sample images of the benign and malignant slices

along with dataset. Finally, retrieve 7328 CT images. Sample images of the benign and malignant slices are depicted in Fig. 3.

a. **Experimental setup**

The performance of the proposed work is measured with the following statistical measures.

$$\text{Accuracy} = \text{\# of images correctly predicted/total \# of images}$$

	Predicted images	
True images	TP	FP
	FN	TN

$$\text{Accuracy} = \frac{TN + TP}{TP + TN + FP + FN} \qquad (7)$$

Here,

TP Predict true-positive images correctly.
TN Predict true-negative images correctly.
FP Predict false-positive images incorrectly.
FN Predict false-negative images incorrectly.

Table 1 Performance measures of the proposed approach with various kernels of SVM

Kernel	Precision	Recall	F1 score	Accuracy
Linear	85	100	92	84.45
Polynomial	87	100	93	86.53
RBF	88	100	93	88.76

The proposed work is applied to classifying lung tumors by using different types of SVM kernels. The dataset splits into training and testing sets; the training set holds 70% of data, and the test set holds 30% of data randomly. The entire dataset images are segmented by using the FCM segmentation algorithm. The segmented images are fed to the LBP algorithm to extract the features. After that, the extracted features are classified by using the SVM kernels such as linear, polynomial, and radial basis function (RBF). 84.45% accuracy is received while classifying with SVM linear kernel. By using polynomial SVM, the proposed work achieved 86.53% accuracy. RBF kernel achieved 88.76% accuracy (Table 1).

5 Conclusion

In this paper, the proposed method efficiently classifies the lung tumors using the SVM kernel. The existing approaches [2–4] were supports for fewer data samples but this approach tested on a large LIDC dataset. This model helps the experts to predict the correct decisions while diagnosing the disease about lung cancer.

References

1. www.wcrf.org, https://www.wcrf.org/dietandcancer/cancer-trends/worldwide-cancer-data. Last Accessed on 01 July 2019
2. M. Alsallal, M.S. Sharif, B. Hadi, R. Albadry, Decision support detection system for lung nodule abnormalities based on machine learning algorithms. J. Contemp. Med. Sci. **5**(3), 165–169 (2019)
3. S. Saraswathi, L.M.I. Sheela, Detection of juxtapleural nodules in lung cancer cases using an optimal critical point selection algorithm. Asian Pac. J. Cancer Prev. (APJCP) **18**(11), 31–43 (2017)
4. R.M. Devarapalli, H.K. Kalluri, V. Dondeti, Lung cancer detection of CT lung images. Int. J. Recent Technol. Eng. (IJRTE) **7**(5S4), 413–416 (2019)
5. Z.H. Deng, H.H. Qiao, Q. Song, L. Gao, A complex network community detection algorithm based on label propagation and fuzzy C-means. Phys. A Stat. Mech. Appl. **519**, 217–226 (2019)
6. T.K. Sajja, M.D. Retz, H.K. Kalluri, Lung cancer detection based on CT scan images by using deep transfer learning. Traitement du Signal (2019)
7. S. Nigam, R. Singh, A.K. Misra, Local binary patterns based facial expression recognition for efficient smart applications, in *Security in Smart Cities: Models, Applications, and Challenges* (Springer, Cham, 2019), pp. 297–322

8. V.N.T. Le, B. Apopei, K. Alameh, Effective plant discrimination based on the combination of local binary pattern operators and multiclass support vector machine methods. Inf. Process. Agric. **6**(1), 116–131 (2019)
9. LIDC-IDRI database, https://wiki.cancerimagingarchive.net/display/Public/LIDC-IDRI. Last Accessed on 01 July 2019

Smart Technologies for Data-Driven Pipeline Risk Assessment

Jason Levy

1 Introduction

This paper describes the design and integration of platforms, architectures, systems, and geospatial tools for the real-time monitoring and alerts of industrial accidents and incidents, with an emphasis on pipeline monitoring and management. Pipelines have become an efficient means of transporting liquid and gaseous materials particularly crude oil, petroleum products, and natural gas for long distances over land, underground, and sub-sea from production locations to the market. However, a ruptured pipeline has the potential of causing serious environmental damage since most of the products being transported are environmentally hazardous in the event of a spillage [29]. Oil and gas infrastructures have been vandalized in many countries including Canada, Columbia, Georgia, Iraq, Mexico, Mozambique, Nigeria, Pakistan, Sudan, Turkey, and USA. The Transportation Safety Board of Canada (TSB) listed recent gas line accidents' data. Figure 1 presents the number of pipeline accidents and their attribute accident rates (per exajoule) from 2005 to 2014 published by the Transportation Safety Board of Canada [50].

In February 2011, an explosion from a natural gas pipeline killed five people in Allentown, Pennsylvania, in September, a 30-in-diameter pipeline in San Bruno, California, exploded killing eight people and burning down three dozen houses, and in July 2011 a pipeline in Kalamazoo, Michigan, ruptured and more than a million gallons of corrosive bitumen from the Canadian tar sands leaked [54]. According to Zeiss [54], aging and pipeline fabrication flaws in the 1950s may have helped to lay the groundwork for these incidents and disasters. More recent problems appear to involve a regulatory, human factors, and political dimension; specifically, unregulated pipeline construction, human error, and information technology challenges appear to be a major culprit. Other literature emphasizes an integrated approach to

J. Levy (✉)
University of Hawaii, Honolulu, HI, USA
e-mail: jlevy@hawaii.edu

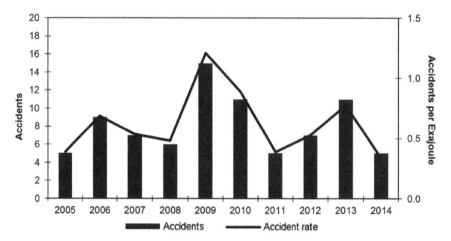

Fig. 1 Number of accidents and accident rate (accidents per exajoule). *Source* The Transportation Safety Board of Canada [50]

understanding pipeline accidents, wherein a confluence of factors, including organizational/social dimensions, human factors, and technologic/engineering failures leads to disaster in these complex systems, particularly when decision makers must take real-time actions under conditions of uncertainty.

In the USA, the Pipeline and Hazardous Materials Safety Administration (PHMSA), a Department of Transportation (DOT) agency, is responsible for developing and enforcing regulations for the safe, reliable, and environmentally sound operation of over a million miles of pipeline. This dedicated organization is responsible for critical health and safety issues relating to hazardous materials transportation. The Norman Y. Mineta Research and Special Programs Improvement Act of 2004 provided a focused research organization with a separately operating administration for pipeline safety and safety. PHMSA provided an opportunity to support US President George W. Bush's 'Management Agenda' initiatives and best information practices.

Many experts believe that such a dedicated administration will enhance the safety and security of the pipeline transportation network. The PHMSA is responsible for nearly one million daily shipments of hazardous materials within the USA by land, sea, and air. This office is responsible for promoting the safe and environmentally sound operation of natural gas and hazardous liquid pipeline systems. Judicious oversight of the US pipeline infrastructure is essential since it accounts for more than half of the energy commodities used in the USA. Pipeline safety regulations and enforcement are essential for supporting relevant oil and gas safety legislation. Established in 2004, the PHMSA enforces the following pieces of legislation:

- Natural Gas Pipeline Safety Act of 1968 (P.L. 90-481)
- Hazardous Liquid Pipeline Act of 1979 (P.L. 96-129)
- Pipeline Safety Improvement Act of 2002

- Pipeline Inspection, Protection, Safety, and Enforcement Act (PIPES) Act of 200
- Pipeline Safety, Regulatory Certainty, and Job Creation Act of 2011 (P.L. 112).

It should be noted that the PHMSA also has a critical responsibility for enforcing related regulations (49 CFR Parts 190–199) and other statutes. Many of these acts were passed as reactive measures after grave crises affected critical infrastructure systems and the natural environment. For example, in response to a double explosion in downtown Richmond, Indiana, the Natural Gas Pipeline Safety Act of 1968 was enacted. The double explosion occurred on April 6, 1968, and killed 41 people and injured more than 150 people. Natural gas leaking from faulty transmission lines was the cause of the primary explosion. On the other hand, the secondary explosion was caused by gunpowder stored inside the building. Twenty buildings in and adjacent to the site of the explosion were damaged (and permanently condemned) as a direct result of the explosions.

The National Gas Pipeline Safety ACT 1968 covers more than 3000 pipeline operators with over 1.6 million miles of gas and hazardous liquid pipelines [17]. From 1994 to 2013, there were more than seven hundred serious incidents in the USA involving gas distribution (causing 278 fatalities and 1059 injuries, with over one hundred million in property damage). In addition, there were more than one hundred serious incidents related to the transmission of gas. This resulted in more than forty fatalities and more than one hundred and ninety injuries. These incidents resulted in half a billion dollars in property damage. A recent Wall Street Journal review found that from the period 2010 to 2013, there were 1400 pipeline spills and accidents in the USA. Moreover, the study found that four-fifths of accidents were discovered by local residents rather than the pipeline owners. Following a cluster of gas pipeline accidents in Bellingham, Washington (State), and El Paso, Texas, decisive action was finally taken by the US government. Both incidents caused numerous fatalities. As a result of these incidents, the federal government decided to improve the quality of geospatial underground pipeline data.

This paper provides insights into the design and integration of multisensor systems and geospatial tools for real-time, event-driven monitoring and management of pipeline incidents and disasters. Section two discusses the motivation of the project and discusses the current available methods for pipeline hazard modeling, monitoring, and alert systems (and other industrial alert and monitoring systems). The remaining parts of the paper are organized as follows: Section three provides an original real-time proposed pipeline monitoring and incident management system architecture and design. This section includes key insights and discussions. Finally, conclusions and future work are found in Sect. four.

2 Importance of the Study and Review of Current Technologies

There are many technologies available for monitoring oil and gas transition processes and failures in pipeline systems. The latest sensor developments incorporate a new generation of high-resolution commercial satellites. A critical function of pipeline monitoring systems is to provide early warnings and alerts of pipeline incidents and possible spills. This role is growing in importance. Remote sensing with satellite or airborne and UAV-based monitoring are practical systems for examining a large area for leak detection and environmental conditions along at-risk pipelines. Leak detection sensors are often placed at key and sensitive locations near the pipeline. These systems are capable of sending telemetry location-based signals. This is often done to improve disaster risk reduction, risk management, and oil spill early warning [48]. Remote sensing techniques continue to play an extremely valuable role in pipeline monitoring and oil spill risk management. They can be used to not only detect the environmental change and identify hazardous areas that are inaccessible [15, 45, 52] but also to reduce costs and improve system performance. It is important to leverage the latest advances in sensor technologies to improve pipeline management. In particular, the next-generation satellites hold tremendous promise to improve the monitoring and early warning of pipeline critical incidents. In particular, satellites are capable of improving the accuracy of a wide range of attributes of relevance to pipeline monitoring (including spatial, spectral, and temporal attributes)

In Fig. 1, the preferences and opinions of stakeholders interested in pipeline risk management issues are considered. Each stakeholder participated in a number of research focus groups and workshops to share their opinions about ways to reduce pipeline risks. The decision makers also participated in the focus groups in order to share their opinions and values. Each stakeholder considered a number of commercial imagery products and pipeline management technologies ranging from airborne and satellite radar to LIDAR, multispectral sensors, and hyperspectral sensors [10, 11, 12, 13, 22, 36, 46]. A wide range of pipeline decision makers participated in the pipeline workshops and focus groups ranging from pipeline engineers and petroleum scientists to non-governmental organizations and social activists. Other focus group participants included private sector firms representing the oil and gas industry, as well as environmental groups. The results of the workshop surveys are shown in Fig. 2.

The use of geomatics solutions for industrial system monitoring has become increasingly important. For historic reasons, the development of satellites was often related to military objectives and applications [44]. There are a wide range of unmanned aerial vehicle (UAV) systems that are capable of providing valuable systems for high-resolution remote sensing and risk management. In recent years, there has been an explosion in the research and development of UAV applications for pipeline risk management with a particular focus on pipeline monitoring, inspection, and crisis management. A large literature discusses the development of processing

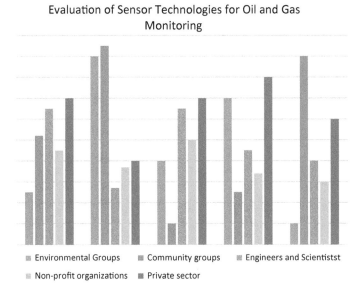

Fig. 2 Evaluation of sensor technologies for oil and gas monitoring by environmental groups, community groups, nonprofit organizations, engineers and scientists, nonprofit organizations, and the private sector

systems for big data as they pertain to pipeline management technologies supervisory control and data acquisition (SCADA). There are also new technologies related to the application of fiber optic solutions for pipeline monitoring. These systems are increasingly used to monitor, control, detect, identify, and localize major oil and gas pipeline anomalies. These risks include pipeline cracks and other damages to oil and gas critical infrastructure. In summary, previous developments in geomatics engineering of various devices and technologies systems and platforms were primarily motivated by the defense sector. However, UAVs and other technologies are now increasingly used to improve the management of industrial challenges such as monitoring oil and gas pipelines. Femi discusses the use of SCADA and fiber optics systems for the monitoring of oil and gas infrastructure. These solutions are capable of providing non-intrusive digital pipeline analysis, management, control, and monitoring. In this way, oil and gas operators can act decisively in the event of a pipeline intrusion. Folga discusses the ability to detect leaks using a variety of leading edge statistical and technologic approaches. In particular, inferential methods can be used to determine leaks by measuring and comparing the quantities of oil present in the pipelines at various locations. Samberg [46] and Folga provide more details.

There is a large body of leading edge research that uses location-based signals to improve oil and gas spill management. These technologies have reached a critical inflection point where they are now capable of providing timely and effective oil spill early warning. There are a wide range of diverse geospatial solutions that have been developed to support pipeline disaster risk management and oil spill critical incident emergency response. These systems include Aloha, SMIS [42], and IDOR^{2D} [51]. There are diverse arrays of timely and important existing models that are highly compatible with GIS systems, including the following tools:

- Oil Spill Information System (OSIS) [35],
- Oil Spill Risk Analysis (OSRA) model [27],
- General NOAA Operational Modeling Environment system (GNOME) [6, 7, 39],
- Dense Gas Dispersion Model (DEGADIS) [9] and
- Atmospheric Dispersion Model for Denser-Than-Air Releases (SLAB) [9, 19].

Although geomatics technologies are continuously improving, incident management is still largely based on paper-driven, disconnected communications, and manual processes [14]. These legacy systems should be improved in order to meet the robust needs of emergency managers in real-world disasters. As reported in Sider [47], technology today plays a modest role in managing oil and gas spills. Historically, pipeline spills, ruptures, and leaks were manually discovered and reported. For example, Sider [47] examines four years of oil spill accident records to conclude that human observations (in the field) account for a majority of pipeline incidents. Clearly, there is a need for improved monitoring technologies.

Recent pipeline accidents and oil spill releases highlight the need for improved pipeline accident monitoring systems. For example, on July 20, 2016, the Husky Energy pipeline release caused 250,000-l of oil to spill in the western part of the Canadian province of Saskatchewan. Husky Energy noted that the leak was immediately detected. The spill caused oil and chemicals to flow downstream toward other communities. The oil spill posed a grave risk to the water supplies of nearby cities. As a result of this disaster, a state of emergency was declared in several areas. More details are provided in Husky Energy's revised incident report [16]. The Husky Energy is instructive for a number of reasons. First, the company conducts monthly tests pipeline looking for flaws. Second, the company inspects the pipeline every two years. Third, despite these inspection processes, clearly more effective monitoring systems are needed.

Similar disasters have occurred in other jurisdictions and regions across North America and Europe. For example, consider the 2010 Enbridge spill crisis in the US state of Michigan. This disaster led to further discussions and analyses pertaining to pipeline safety and the reliability of leak detection systems [23]. It has been shown that the type of organization responsible for disaster monitoring—and the type of leak detection decision support system used—is important for a number of reasons. Most importantly, private sector oil and gas companies using computerized oil spill detection systems are less likely to identify oil spills than governmental agencies, community groups, nonprofit organizations, private citizens, and/or emergency responders. More discussions about these issues are found in a report commissioned by the Pipeline and Hazardous Materials Safety Administration [23].

Pipeline monitoring and incident management systems typically involve the real-time remote sensing of pipeline situations. There is a large literature focusing on information dissemination in the event of a pipeline disaster. Other critical issues involve the integration and analysis of oil and gas spill data from various sources including emergency management organizations, weather agencies, environmental monitoring groups, geospatial scientists, and others. Real-time remote sensing and communication technologies have made significant progress in recent years, particularly as they relate to pipeline incident detection and response. There has been an increased focus on the real-time monitoring, detection, and management of pipelines and other critical infrastructure. Unmanned aerial systems (UAS) have been increasingly used for operational oil spill surveillance, monitoring, and assessment. Small and medium-sized UAS are typically used for incident detection and response. These technologies are extremely valuable for regions with limited accessibility. The literature shows that this offers a cost-effective alternative that reduces long-term monitoring expenses and human risks [26, 40]. A growing number of oil and gas companies are now focusing on unmanned aerial systems with application to disaster risk reduction and oil and gas pipeline inspection, surveillance, and monitoring [1]. For example, governmental agencies, including the U.S. Coast Guard and National Oceanic and Atmospheric Agency, have incorporated UAS into oil spill response exercises and test applications. Indeed, aerial monitoring is recognized and increasingly adopted in the petroleum industry as a desired and cost-effective approach to acquire crucial pipeline integrity data [34]. Kim et al [33] explore a robot agent-based technology with mobile sensors to promote active and corrective pipeline monitoring and maintenance. Yan and Chyan [53] use distributed fiber optic sensor systems in order to simultaneously measure pipeline temperature and strain. Wireless sensor networks (WSNs) have been widely applied for oil and gas pipeline monitoring: Key benefits include low cost, ease of deployment, and ability to cater for data acquisition at great spatial and temporal scales [28, 41, 49].

While real-time integration and pipeline analysis is critical for event-driven decision making, advances in remote sensing technologies have led to new challenges related to the theory and practice of oil spill management. Accordingly, it is important to re-examine theories, concepts, models, and practices pertaining to spatial pipeline monitoring decision support. A key challenge relates to scalability. It is essential to ensure that the monitoring system is independent of pipeline characteristics (e.g., shape, size) as noted by a number of scholars [32, 40]. Adaptable and extensible solutions are also essential for the integration of data from a variety of sources. These issues become increasingly critical as the number of systems and the complexity of each system increase. System adaptability and extensibility are also essential to cope with the changes to the oil spill risk management system with minimal effort [18, 25].

However, traditional oil spill interaction models and decision support systems are based on the request/reply communication approaches in which clients–servers passively respond to requests [21, 31]. Request/reply models may overload systems leading to a total collapse [24]. Traditional request/reply interaction models and systems may also be unable to support a growing number of clients and may be unable to adequately track information needs from various data sources [37, 43]. The latest research involves creative and timely cloud-based and web-driven applications that use a distributed environment to improve communication, share information, and promote the real-time delivery of dynamic events [5, 20]. It is important to capitalize on advances in sensor technologies.

A wide range of decision support systems for pipeline spill assessment, monitoring, management, and incident management have been developed [6, 8, 35]. However, many of these incident management systems are stand-alone off-line software packages which fail to holistically integrate advanced sensors for real-time monitoring in a coherent and systematic fashion. As a result, these legacy systems have limited effectiveness: They have limited functionalities that cover only a small fraction of pipeline hazards and fail to properly take into account all phases of comprehensive, all-hazards pipeline incident/emergency management: prevention, mitigation, response, recovery, and preparedness. Moreover, such legacy systems typically possess a unique database system which makes it difficult to integrate external systems with other important risk assessment and disaster management functionalities.

3 Advances in Pipeline Monitoring and Incident Management System Design

An important contribution of this research is the design and integration of a comprehensive system for oil and gas pipeline monitoring and incident management that holistically integrates a large number of components. The first component is an operational real-time monitoring and alert system. This involves the ability to provide rapid and high-quality monitoring. Here, a critical technology involves telemetry sensors and early warning. The second technology involves geospatial mining tools and model tools. These components are essential to increase the accuracy of data storage and to improve big data processing and analysis. The third component involves the use of telecommunication systems and infrastructure. This is essential to enhance information dissemination, data sharing, and multimedia analyses. The fourth component involves a command and control system infrastructure that can be used to improve the command, control, administration, and coordination of crisis response.

3.1 Previous System Development

A large number of systems have been developed to monitor a diverse range of natural, health related, and technologic hazards ranging from tsunamis, storm surges, forest fires, and hazardous materials incidents [30]. One important field involves large-scale, cross-boundary oil spills, and urban fire risks [2]. This contribution builds on previous incident management systems that have proven valuable disaster risk reduction, and systems for hazard analysis and command and control [38]. Enterprise messaging systems constitute a key technology for oil spill management. Key technologies involve the use of real-time, effective, and advanced GIS-based systems for disaster notification and alerts [3, 4, 30]. Publish and subscribe systems have been developed for tornado and hurricane risk management that improved geospatial-based disaster management, situational awareness, and crisis response. Original geospatial-based interaction frameworks have been developed to improve situational awareness in the natural hazard field (such as coastal storm risks). We have designed many oil spill management systems that integrate dynamic geospatial information sources and incorporate leading edge event-driven GIS systems. These systems allow for a large number of clients and decision makers to use the oil spill management system.

To improve oil spill disaster risk management, our research team has developed novel and powerful oil spill detection and monitoring systems that include a wide range of leading edge technologies and systems. Key technologies and research innovations include the following: middleware systems that use a publish–subscribe interaction framework for oil spill management; the real-time processing of geospatial oil spill events and transactions (with a focus on spatially related oil pipeline information of interest); the use of scanning laser environmental airborne fluorosensor (SLEAF) technologies; web-based oil spill emergency management solutions that allow agencies and firms to determine the location of trajectory of oil spills; as well as laser fluorosensor data processing. It is recognized that clients have heterogeneous and specialized data interests.

A key question involves the best way to improve the flow of information in real time. To achieve this objective, all spatial oil spill events and services can be accessed in real time using the enterprise service bus (ESB) and the TIBCO enterprise message service. Taken together, these advances in web-based geomatics and knowledge engineering provide critical technologies to improve oil spill crisis management, environmental risk analysis, and emergency response and recovery. We have also incorporated a wide range of image processing systems to detect oil spills. In summary, our oil spill disaster risk management research incorporates new technologies in database management systems in order to provide effective disaster risk reduction for the oil and gas sector powered by reliable geospatial data processing systems.

3.2 System Architecture

The event-based architecture ensures timely interactions among all systems (including hardware, sensors, and pipeline critical infrastructures) within a wide-scale network using event-based communication technologies. This flexible system allows for critical information to be rapidly shared with an unlimited number of heterogeneous clients and subscribers ranging from civil engineering academics and environment agencies to industrial control centers, regulatory governmental bodies, pipeline operators, and weather providers. Advances in GIS technologies are leveraged to develop an intelligent information communication infrastructure that fosters robust disaster risk reduction for the oil and gas sector. As shown in Fig. 3, major components of the pipeline monitoring and management system include the following systems, models, and services: Web-based notification and data transaction services; decision support

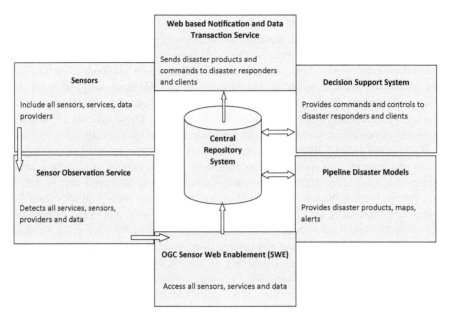

Fig. 3 Major components of the pipeline monitoring and management system: Web-based notification and data transaction services; decision support systems; sensor systems; sensor observation services; sensor web enablement (SWE); and pipeline disaster models

systems; sensor systems; sensor observation services; sensor web enablement (SWE); and pipeline disaster models. Figure 4 presents the overall system architecture while Fig. 5 shows the configuration and flow of data. More details about the integrated pipeline incident management, operation, and administration system are provided in the text that follows.

The enterprise service bus (ESB) involves the use of a topic-based event-driven message middleware. The oil and gas spill command and control system improves oil and gas pipeline incident risk management using the system portal and ESB. There are a wide range of clients employed in this robust oil spill management system: One client in the system is responsible for pipeline reporting: this key task involves reporting a potential oil spill and communicating the information with relevant environmental agencies. A second client is responsible for logistics and facility management. Other clients carry out tasks related to oil spill recovery operations, budgeting, and administration. More information about related publish/subscribe systems and the enterprise service bus (ESB) in the communication layer can be found in Assilzadeh et al. [3, 4] and Kassab [31]. There are a number of outputs, modeling products, and functionalities of the proposed command and control system including the thematic disaster maps, incident reports, document management systems, and fillable disaster forms.

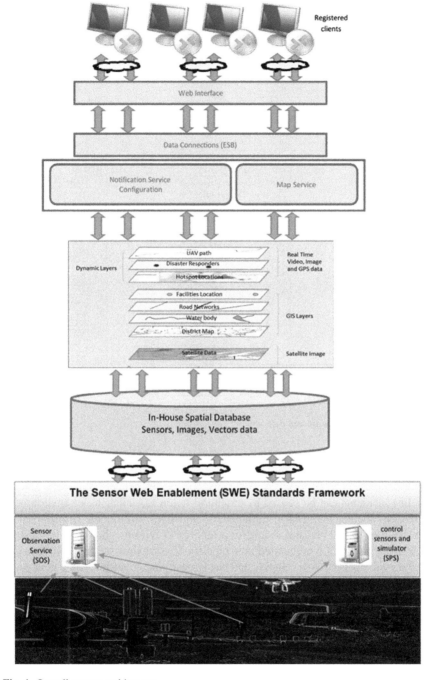

Fig. 4 Overall system architecture

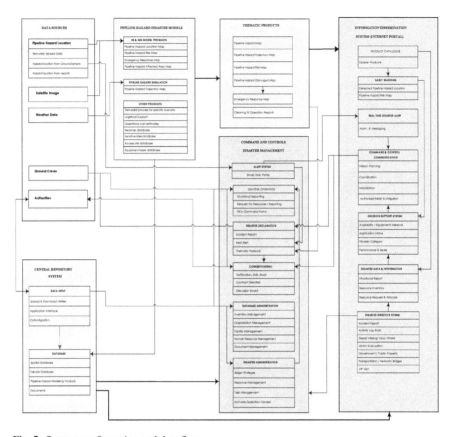

Fig. 5 System configuration and data flow

4 Conclusions

Pipeline incidents can be caused by structural failures, vandalism, or environmental hazards such as hurricanes, earthquakes, landslides, and floods. Disaster risk is increasing in the oil and gas sector for a number of reasons: the aging of existing oil and gas pipelines, unregulated new pipeline construction, and the relocation of oil and gas industries to developing and newly emerging countries where policy and regulatory institutions are weak. As a result of these changes, the potential for significant pipeline accidents and crises is increasing each year. The consequences of these pipeline disasters are damaging to society, the economy, and the environment. Accordingly, a comprehensive and integrated pipeline monitoring and management Internet portal system is designed which contains four major components: an early warning and alert system, geospatial data mining tools and model components, a communication system, and a command and control system.

This paper describes the design and integration of leading edge platforms, architectures, systems, and geospatial tools for the real-time monitoring and management of pipeline hazards. A unique geospatial design is put forth for oil and gas incidents involving pipeline transportation systems. Threats to both the land and aquatic environment are considered. A comprehensive review of the existing geospatial pipeline management systems and architectures is carried out. A multisensor solution for detecting pipeline fractures and failures is put forth. Geospatial tools and models for data processing and analysis are shown to provide quality early warning and critical information for real-time oil spill response. The proposed system architecture supports real-time geospatial disaster management and decision making and integrates the use of thematic maps and incident reports. An emphasis is placed on compressive, integrated, and all-hazards crisis management. A command and control module is designed to provide disaster management response and to coordinate incident management tasks. An integrated event-based system is developed so that components of the system seamlessly communicate and clients request and receive necessary disaster information. The proposed geomatics engineering approach presents an effective solution to the growing problem of pipeline accidents by incorporating both telemetry and Internet-based communication infrastructure. It is shown that the system architecture has the potential to reduce not only emergency response times but also the impacts of pipeline incidents in the context of the broader environmental, regulatory, socioeconomic, and physical context.

The implemented design provides timely and innovative solutions that expand upon isolated systems currently used for pipeline incident management by integrating them into a holistic and powerful multi-functional pipeline incident management system. The system covers all aspects of pipeline incident management with an emphasis on pipeline failure early warning, environmental monitoring, disaster simulations, as well as incident monitoring and emergency response and recovery. In particular, the developed system improves the real-time operation and management of pipelines. As a result, possible incidents become more transparent, and disaster response is more efficient. It has been shown that the proposed system protects the environment, saves money, and improves response efficiency. Existing modules for pipeline accident modeling, situational awareness, monitoring, and simulation are incorporated into the real-time geospatial architecture. This original system improves communication between deployed sensors and devices on either mobile or fixed platforms.

Pipeline monitoring and management systems continue to suffer from a lack of reliable and scalable architectures that allow for the real-time delivery of critical information and dynamic events. This is because existing incident management systems are often stand-alone off-line software packages which fail to holistically integrate advances in sensor technologies for real-time monitoring in a robust, integrated, coherent, and systematic fashion. Such systems also have limited functionalities that cover only a small fraction of pipeline hazards and threats. Existing systems also fail to properly take into account all phases of comprehensive pipeline incident/emergency management: prevention, mitigation, response, recovery, and preparedness. Moreover, it is often difficult to integrate external systems into the existing pipeline management systems due to challenges with legacy database systems. In summary, we highlight that current systems are incomplete, disjointed, and possess rigid database formats and problematic system configurations which limit their effectiveness. The herein proposed system improves risk assessment and disaster management by including robust functionalities that are essential for comprehensive and integrated pipeline incident management.

References

1. J. Allen, B. Walsh, Enhanced oil spill surveillance, detection and monitoring through the applied technology of unmanned air systems. Int. Oil Spill Conf. Proc. **2008**(1), 113–120 (2008)
2. H. Assilzadeh, W.J. de Groot, J.K. Levy, Spatial modelling of wildland fire danger for risk analysis and conflict resolution in Malaysia: linking Fire Danger Rating Systems (FDRS) with Wildfire Threat Rating Systems (WTRS). Geocarto Int. J. **27**, 4 (2012)
3. H. Assilzadeh, Z. Zhong, A.S. Kassab, Y. Gao, J.K. Levy, Development of an even-driven and scalable oil spill monitoring and management system, in Canadian Geomatics Conference and ISPRS Technical Commission I Symposium 2010. Calgary, Canada, 14–18 June 2010
4. H. Assilzadeh, Z. Zhong, A.S. Kassab, Y. Gao, J.K. Levy, Development of an even-driven and scalable oil spill monitoring and management system, in *Proceedings of Canadian Geomatics Conference and ISPRS Technical Commission I Symposium*. Calgary, Canada, 2010
5. J. Bartley, R. Narayanan, D. Cordes, ArcGIS server performance and scalability—optimizing GIS services, in *Technical Workshops, Proceedings of Esri User Conference*. San Diego, CA, 20–24 July 2015
6. C.J. Beegle-Krause, General NOAA oil modeling environment (GNOME): a new spill trajectory model, in *Proceedings of IOSC*, vol. 2 (Mira Digital Publishing Inc., Tampa, FL, St. Louis, MO, 2001), pp. 865–871
7. C.J. Beegle-Krause, C. O'Connor, *GNOME Data Formats and Associated Example Data Files* (NOAA Office of Response and Restoration, Emergency Response Division, Seattle) (formerly Hazardous Materials Response Division), pp. 49, 2005
8. C.J. Beegle-Krause, C. O'Connor, G. Watabayashi, I. Zelo, C. Childs, NOAA safe seas exercise 2006: new data streams, data communication and forecasting capabilities for spill forecasting, in *Proceedings of AMOP. Edmonton*, Alberta, Canada, 2007
9. D.N. Blewitt, J.F. Yohn, D.L. Ermak, An evaluation of SLAB and DEGADIS heavy gas dispersion models using the HF spill test data, in *Proceedings of the AIChE International Conference on Vapor Cloud Modeling*. Boston, MA, 1987
10. C. Brekke, A. Solberg, Oil spill detection by satellite remote sensing. Remote Sens. Environ. **95**, 1–13 (2005)
11. C. Brown, M. Fingas, Review of oil spill remote sensing. Spill Sci. Technol. Bull. **4**, 199–208 (1997)
12. C. Brown, M. Fingas, Review of the development of laser fluorosensors for oil spill application. Mar. Pollut. Bull. **47**, 477–484 (2003)
13. C. Brown, M. Fingas, R. Hawkins, Synthetic aperture radar sensors: viable for marine oil spill response, in *Proceedings of the 26th Arctic and Marine Oil Spill Program (AMOP) Technical Seminar*. Victoria, Canada, 2003
14. J. Clark, A 21st century response: implementing GIS-based incident management, in *Paper Sessions, Proceedings of ESRI User Conference*. San Diego, CA, 20–24 July 2015
15. S. Claudia, W. Urs, in *Investigation of ERS SAR Data of the Tandem Mission for Planning and Monitoring of Siberian Pipeline Tracks* (1997). Retrieved Feb 2016 from http://earth.esa.int/workshops/ers97/papers/streck/index.html
16. DNSUnlocker, in *Husky Admits Crews Missed Leak Night of Saskatchewan Oil Spill* (2016). Retrieved Sept 2016 from http://www.cbc.ca/news/canada/saskatchewan/husky-oil-spill-government-july-28-update-1.3699007
17. DOE (Department of Energy), in *Transmission, Distribution and Storage* (2011). Retrieved Feb 2016 from http://fossil.energy.gov/programs/oilgas/delivery/index.html

18. P. Dey, Decision support system for inspection and maintenance: a case study of oil pipelines. IEEE Trans. Eng. Manage. **51**(1), 47–56 (2004)
19. D.L. Ermak, S.T. Chan, Recent developments on the FEM3 and SLAB atmospheric dispersion models, in *Proceedings of the IMA Conference on Stably Stratified Flows and Dense-Gas Dispersion*. Chester, England, 1986
20. ESRI, *Answering Emergency Management Information Needs*, 2012 edn. (ArcUser, 2012)
21. P.T. Eugster, P.A. Felber, R. Guerraoui, A.M. Kermarrec, The many faces of publish/subscribe. ACM Comput Surv (CSUR) **35**, 114–131 (2003)
22. M.F. Fingas, C.E. Brown, J.V. Mullin, A comparison of the utility of airborne oil spill remote sensors and satellite sensors, in *Proceedings, Fifth Conference on Remote Sensing for Marine and Coastal Environments*, vol. 1 (Environmental Research Institute of Michigan, Ann Arbor, MI, 1998), pp. 171–178
23. Flow Control, in *Pipeline Leak Detection Systems Faulty* (2013). Retrieved Sept 2016 from http://www.flowcontrolnetwork.com/pipeline-leak-detection-systems-faulty-report-says/
24. M.J. Franklin, S.B. Zdonik, *"Data in Your Face": Push Technology in Perspective* (ACM, New York, USA, Seattle, Washington, USA, 1998), pp. 516–519
25. T. Gannon, S. Madnick, A. Moulton, M. Sabbouh, M. Siegel, H. Zhu, Framework for the analysis of the adaptability, extensibility, and scalability of semantic information integration and the context mediation approach. Working paper CISL# 2009-02 (2009). http://hdl.handle.net/1721.1/65625
26. D. Hausamann, W. Zirnig, G. Schreier, P. Strobl, Monitoring of gas pipelines—a civil UAV application. Aircr. Eng. Aerosp. Technol. **77**(5), 352–360 (2005)
27. P.M. James, J.R. Walter, M.F. Charles, J. Zhen-Gang, R.B. Gail, Overview of the oil spill risk analysis (OSRA) model for environmental impact assessment. Spill Sci. Technol. Bull. **8**, 529–533 (2003)
28. I. Jawhar, N. Mohamed, K. Shuaib, A framework for pipeline infrastructure monitoring using wireless sensor networks, in *Wireless Telecommunications Symposium, 2007. WTS 2007*, Apr 2007, pp. 1–7
29. Joint UNDP/World Bank Energy Sector Management Assistance Programme (ESMAP), in *Russia Pipeline Oil Spill Study* (2003). Retrieved Feb 2016 from https://www.esmap.org/sites/esmap.org/files/03403RussiaPipelineOilSpillStudyReport.pdf
30. A. Kasab, S. Liang, Y. Gao, Real-time notification and improved situational awareness in fire emergencies using geospatial-based publish/subscribe. Int. J. Appl. Earth Obs. Geoinf. **12**(6), 431–438 (2010)
31. A. Kassab, in *Geospatial-based Publish/Subscribe: Improving Real-time Notification and Situational Awareness in Fire Emergency* (2009). Retrieved Feb 2016 from http://www.geomatics.ucalgary.ca/research/publications
32. J. Kim, G. Sharma, N. Boudriga, S. Iyengar, N. Prabakar, Autonomous pipeline monitoring and maintenance system: a RFID-based approach, in *Lecture Notes in Computer Science*, vol. 8473 (2014)
33. J. Kim, G. Sharmaz, N. Boudrigax, S. Iyengarz, SPAMMS: a sensor-based pipeline autonomous monitoring and maintenance system, in *2010 Second International Conference on Communication Systems and Networks* (*Comsnets*), Jan 2010
34. B. Kramer, *Eye in the Sky* (North American Oil & Gas Pipelines, 2013)
35. M. Leech, M. Walker, M. Wiltshire, A. Tyler, OSIS: a windows 3 oil spill information system, in *Proceedings of 16th Arctic Marine Oil Spill Program Technical Seminar*. Calgary, Canada, 1993, pp. 549–572
36. N.J. Maya, L. Jason, G. Yang, Advances in remote sensing for oil spill disaster management: state-of-the-art sensors technology for oil spill surveillance. Sens. Online J. **8**, 236–255 (2008)
37. G. Muhl, *Generic Constraints for Content-Based Publish/Subscribe* (Springer, London, UK, Trento, Italy, 2001), pp. 211–225
38. M.J. Nand, J.K. Levy, Y. Gao, Advances in remote sensing for oil spill disaster management: state-of-the-art sensors technology for oil spill surveillance. Sensors **8**(1), 236–255 (2008)

39. NOAA, in *General NOAA Oil Modeling Environment (GNOME), User's Manual NOAA Office of Response and Restoration* (Hazardous Materials Response Division, Silver Spring, MD, 2002)
40. J. Ondráček, O. Vaněk, M. Pěchouček, Monitoring oil pipeline infrastructures with multiple unmanned aerial vehicles, in *Lecture Notes in Computer Science*, vol. 8473 (2014), pp. 219–230
41. G. Owojaiye, Y. Sun, Focal design issues affecting the deployment of wireless sensor networks for pipeline monitoring. Ad Hoc Netw. **11**(3), 1237–1253 (2013)
42. H.M. Paul, J.L. Eugene, B.D. Edsel, P.D. James, D.A. Mark, Development of a GIS-based spill management information system. J. Hazard. Mater. **3**, 239–252 (2004)
43. P.R. Pietzuch, Hermes: a scalable event-based middleware. Unpublished thesis. University of Cambridge, 2004
44. F. Remondino, L. Barazzetti, F. Nex, M. Scaioni, D. Sarazzi, UAV Photogrammetry for mapping and 3D modeling—current status and future perspectives, in *International Archives of the Photogrammetry, Remote Sensing and Spatial Information Sciences*, vol. XXXVIII-1/C22 (ISPRS Zurich Workshop, Switzerland, 2011)
45. R. Reuter, H. Wang, R. Willksmm, K. Loquay, A laser fluorosensor from maritime surveillance: measurement of oil spills. EARSeL Adv. Rem. Sens. **3**(3), VII (1995)
46. A. Samberg, Advanced oil pollution detection using an airborne hyperspectral lidar technology, in *Proceedings of SPIE, the International Society for Optical Engineering*, vol. 5791. Laser Radar Technology and Applications X (2005)
47. A. Sider, High-tech monitors often miss oil pipeline leaks. Business, Wall Street J. (2014)
48. I. Stoianov, L. Nachman, S. Madden, *PIPENET: A Wireless Sensor Network for Pipeline Monitoring* (2011). Retrieved Feb 2016 from http://db.csail.mit.edu/pubs/ipsn278-nachman.pdf
49. I. Stoianov, L. Nachman, S. Madden, T. Tokmouline, M. Csail, PIPENET: a wireless sensor network for pipeline monitoring, in *Proceedings of 6th International Symposium on Information Processing in Sensor Networks* (Cambridge, MA, USA, 2007)
50. Transportation Safety Board of Canada, Statistical Summary—Pipeline Occurrences Catalogue No. TU1-11E-PDF. ISSN 1701-6606 (2016). Retrieved Feb 2016 from http://www.tsb.gc.ca
51. I.K. Tsanis, S. Boyle, A 2D hydrodynamic/ pollutant transport GIS model. Adv. Eng. Softw. **32**, 353 (2001)
52. E.R. William, D. Subijoy, in *Oil Spill and Pipeline Condition Assessment Using Remote Sensing and Data Visualization Management Systems* (1997). Retrieved Feb 2016 from http://www.spectir.com/wp-content/uploads/2012/02/Oil_Spill_and_Pipeline_Condition_Assessment_Using_Remote_Sensing_and_Data_Visualization_Management_Systems-Roper.pdf
53. S. Yan, L. Chyan, Performance enhancement of BOTDR fiber optic sensor for oil and gas pipeline monitoring. Opt. Fiber Technol. **16**(2), 100–109 (2010)
54. G. Zeiss, in *Pipeline Accidents and Maintaining High Quality Data About Facilities* (2011). Retrieved Feb 2016 from http://geospatial.blogs.com/geospatial/2011/04/pipeline-explosions-and-data-quality.html

Neuro-fuzzy Knowledge Processing for Smart Emergency Management: Advances in Computational Intelligence and Seismic Community Resilience

Jason Levy

1 Introduction

Disaster and emergency management challenges are complex and ill-structured group decision processes characterized by extreme instability, high decision stakes, large-world uncertainty, severe time constraints, dynamic, value-laden trade-offs and multi-faceted, and self-organizing negotiations [5]. This paper proposes that solutions to these "wicked" [9] national security and disaster management challenges can be best addressed by the latest advances in "discrete knowledge processing systems". A catastrophic disaster event has all the hallmarks of wicked problems, including a lack of a "stopping rule" (when will the disaster end?), a lack of a definitive problem formulation (creating a solution changes our understanding of the problem) and lack of an ultimate test for problem solution (it is impossible to know how the consequences of global terrorism, pandemics and even climate change will play out). As a result, problems involving global catastrophes do not have a well-defined set of possible solutions. Moreover, decision makers have widely divergent views of acceptable emergency management strategies outcomes and may not agree on fundamental issues (i.e., which group is responsible for a specific terrorist attack? How many communities will be affected by an earthquake? Should society prepare for a 1 in a 1000 year flood? What is the range of acceptable future sea level rise? Thinking globally, what will be the direction of future climate change?).

Given a large multidimensional data set and a "wicked" emergency management problem, the normal assumption is a nonlinear model unless there is a priori knowledge that the system is linear. Systems are complex enough not to have closed form nonlinear models that can be derived from analysis using basic underlying science that would have to rely on empirical data to come up with a model. Although there is no guarantee that this model would be closed form (sets of equations of finite length).

J. Levy (✉)
University of Hawaii, Honolulu, HI, USA
e-mail: jlevy@hawaii.edu

In disaster modeling, nonlinear regression is often used to fit a model $y = f(\mathbf{x}, \boldsymbol{\theta}) + \varepsilon$ where f is a nonlinear function and $\mathbf{x}$ and $\boldsymbol{\theta}$ are vectors, i.e., $\mathbf{x} = [x_1, x_2, ..., x_n], \boldsymbol{\theta} = [\theta_1, \theta_2, ..., \theta_n]$ and ε is some noise component. Note that y can be static or dynamic and is scalar (but can be a vector for several f s). If dynamic, $\mathbf{x}$ includes samples of variables at different times, causing y to change with time. While it is possible to reduce disaster management problems that involve complex disaster phenomena and large amounts of multidimensional data to relatively simple systems with few components, this assumes that much of the data are collinear with the state variables of the simpler system and excluded from the study. In practice, statistical analysts often reduce the number of variables in a disaster model to avoid combinatorial explosion, thereby making the problem easier to model with conventional statistical techniques (such as regression analysis, principal component analysis and subtractive clustering). However, this is less an ideal: eliminating variables that influence the behavior of the system under investigation can significantly reduce model accuracy. Moreover, how can one justify excluding potentially valuable information, particularly when thousands of lives may be at stake.

Three basic techniques that do not rely on order reduction and that can thus deal with the resulting combinatorial explosion of (real-world) "wicked" disaster management problems are (a) the use of parallel processing computers (b) Combs unions rule configuration [2] and (c) discrete neuro-fuzzy processing. The third technique is the most recent and is now described and applied to earthquake reconstruction in the city of Bam, Iran.

The unique and powerful concept of discrete knowledge processing is introduced to handle higher order complex (dynamic, nonlinear) systems that characterize disaster problems. Discrete knowledge processing is contrasted to traditional "continuous knowledge processing systems" used in traditional soft systems, including neural networks (NN), fuzzy-encoded multilayer perceptrons (MLPs), ANFIS (adaptive network-based fuzzy inference systems, a fuzzy inference system implemented in the framework of adaptive networks) and radial basis function networks (RBFNs), which have been shown to be equivalent to zero-order Sugeno fuzzy systems [3].

Discrete knowledge processing systems are shown to be more efficient than continuous knowledge processing for complex disaster problems. This is because discrete knowledge models not only allow us to model data complexity (large numbers of data columns) but offers the possibility of knowledge discovery. That is the discovery of rules previously unknown indicated by the clustering of sparse array components after training with data. This enables us to more effectively reason about the behavior of the disaster system. In this paper, training and validations of data sets will be used to establish the relationships among key disaster variables using the discrete knowledge processing approach.

The paper is organized as follows. Section 2 discusses Combs proposed "union rule configuration" to improve the efficiency of making inferences making in high-order fuzzy models. Next, Sect. 3 describes the representation and mathematical formulation of discrete knowledge processing. Section 4 provides illustrative numerical examples. Specifically, an earthquake reconstruction case study in Bam, Iran illustrates data analysis and processing with the proposed discrete neuro-fuzzy technique.

Training errors, error checking and the clustering of rules are discussed in Sect. 5, and some of their important aspects are presented. Rules are found if extracts cluster in the sparse array which can be easily demonstrated graphically for low-order problems. Finally, Sect. 6 concludes the work providing also suggestions for future important directions and extensions.

2 Parallel Processing and Combs Method

Combinatorial explosion translates into many rules, and parallel processing is a natural solution where several rules are evaluated in parallel on separate processors. The problem is the scaling of such systems. The term 'explosion' as a "curse of dimensionality" (coined by Bellman [1] is very appropriate for disaster management problems [8]. For example, an emergency manager applying a neuro-fuzzy system with the minimum of two membership functions (MFs) per variable and fifty disaster variables has one trillion times the number of rules as one with an equivalent number of MFs per variable and 10 variables ($2^{50}/2^{10}$) when grid partitioned. It would require one trillion processors running in parallel to give the same performance as that of a ten variable problem. So, unless there is an equivalent explosion in parallel processing which not even Moore's law (a law that describes trends in computing hardware) provides for, then there clearly are limitations.

Combs proposed a "union rule configuration" to improve the efficiency of inference making in high-order fuzzy models. The principles of Combs method are rather simple in retrospect. The rules that are said to be explosive in numbers are as a result of the conjunction of antecedents called an intersection rule configuration (IRC). Combs indicated that such rules have an equivalent union rule configuration (URC) which involves the disjunction of antecedents. Processing such rules in URC form is said to be more efficient than in IRC form. One complication is that a URC rule that models a given IRC rule also generates other spurious IRC rules thus the need for corrective terms to be introduced later introduced. No mechanism for regression of these URC rules directly from data was disclosed, so the assumption is that conventional rules (which can be explosive in number) must first be obtained and then converted into URC form for more efficient processing. So, even if the spurious generation of IRC rules and corrective terms is manageable, the need to resort to the same IRC rules for regression is counterproductive.

3 Discrete Neuro-fuzzy Models

The discrete neuro-fuzzy technique involves challenging the geometric model of rules and how they fire. Even if there is an explosion in the total number of rules due to complexity at any point in time for a given state, only a few rules apply. The conventional geometric model of the firing of a rule assumes a vector field surrounding the rule. The closer to the center of the rule the stronger the field (the more the rule fires) and the further away the weaker the field. This model is used whether the dimensionality of the vector is low or high. An inference system is a collection of such rules, i.e., a set of vector fields. Which rules apply can only be determined by calculating the field strength of all rules exhaustively (even if there are an explosive number of such rules). The fact that one rule firing (high field strength) automatically means other rules not firing (low field strength) cannot be exploited by this model and can only be discovered after the fact with an exhaustive search. That is even those rules that in retrospect are known not to be relevant must be evaluated which can be very numerous.

3.1 Introduction to the Discrete Neuro-fuzzy Technique

Rather than multiple independent vector fields, the discrete neuro-fuzzy approach uses a tensor field model of a fuzzy inference system (see Fig. 1). A tensor field is a variable geometric quantity in m-dimensional space represented by an array with n-indices indicating its rank. The simplest geometric quantity is a scalar which is a tensor of rank zero (no indices) followed by a vector (related collection of scalars) which is a tensor of rank one (one index) then a matrix (related collection of vectors) which has two indices (tensor of rank two) followed by higher ranked tensors (related collection of lower ranked tensors). High-rank tensor fields are modeled using sparse multidimensional arrays. The benefit of the tensor filed model is that even though large numbers of rules are encoded in the multidimensional array only the few that are relevant for a given state need to be evaluated without requiring an exhaustive search (Fig. 1).

Particularly in Fig. 1, an input vector takes the path shown and the firing strengths of rules as the input vector changes are given as vectors. The locations of rules are also shown in Fig. 1. The generalized bell membership function indicates typical firing strengths as input vectors (positions on the curve) change. These increase as the rule center is approached. In the conventional model, rule firing strengths need to be calculated exhaustively for each rule. This is inescapable even though it is clear that if rule 1 has a higher firing strength than rule 2 then the latter need not be calculated since it would be automatically low (which can be approximated to zero). The independent vector field model used in conventional neuro-fuzzy systems is incapable of exploiting this interrelationship.

Figure 1c is called the tensor field model because a higher order geometric quantity is associated with each point in the field as shown by the little triangles. As can be seen, the same information is encoded from both Fig. 1a, b. Furthermore, if for each input vector (or point on the curve) a small "square" (or equivalent box in three dimensions) is drawn around it, then only the firing strengths of rules within this square need be calculated. The larger the size of the square, the more precise the answer becomes. In training of the new model, only parameters of components

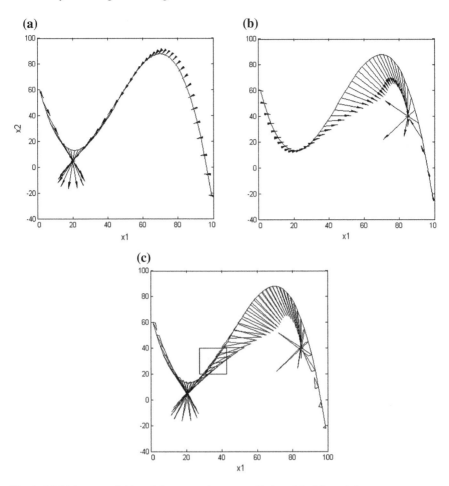

Fig. 1 Multiple vector field model compared to tensor filed model of fuzzy inference system

within this square need be updated and in recall only components within the box need be evaluated.

The advantage only becomes obvious when there are large input vectors (many inputs) resulting in many rules, where the generalized bell membership functions only produce high firing strengths in narrow regions around the rules. In this case, the square becomes a hypercube and the size of the hypercube can be reduced and the number of rules within can be a very small fraction of the total number of rules that would otherwise in the conventional model be evaluated exhaustively. This narrowing of the response of a rule makes it so discrete that it can no longer be considered a rule as one understands it, but it can be a component of a rule, with a cluster of components representing individual rules. This gives rise to the 'discrete' nomenclature for the new model. The descriptions of the model as a tensor field model and discrete model are complimentary. It can be shown that the method can give accurate results equivalent to that produced by conventional neuro-fuzzy methods but with the higher efficiency inherent in the advanced model. There is a large literature dealing with these issues including the use of clustering rules [6], fuzzy inference applications [10] and dimension reduction [7].

3.2 Discrete Neuro-fuzzy Tensor Calculus

The tensor field is represented computationally by a sparse multidimensional array. Changes to the field are indicated by changes to components of the array in a neuro-fuzzy tensor calculus. Changes to components in the array are made in order to make the output produced by the system given by Eq. (4) match the target data. This is best achieved by the minimization of the error function of Eq. (1) which allows for a conventional solution to the problem. The overall algorithm is described in Fig. 2 and when extended to the multidimensional problem the "sub-hypercube" (in flowchart of Fig. 2) is equivalent to the "square" in Sect. 3.1. The details of the calculation are performed in Fig. 2.

More details are now provided about the theory of neuro-fuzzy tensor calculus. For m, data points let

$$e(\boldsymbol{\theta}) = \sum_{k=1}^{M} (y(\boldsymbol{\theta})_k - \hat{y}_k)^2, \quad \boldsymbol{\theta} = [\boldsymbol{\theta}_c, \boldsymbol{\theta}_b, \boldsymbol{\theta}_p] = [\mathbf{c}, \mathbf{b}, \mathbf{p}] \quad (1)$$

We now seek to find $\boldsymbol{\theta}^*$ where

$$\boldsymbol{\theta}^* = \arg\min e(\boldsymbol{\theta}) \quad (2)$$

This gives rise to the set of parameters which minimize the error function. Next, we calculate the square of the error function for a single point

$$e_k(\boldsymbol{\theta}) = (y(\boldsymbol{\theta})_k - \hat{y}_k)^2 \quad (3)$$

There are three types of adaptive parameters used in the discrete neuro-fuzzy tensor calculus: component indices $\boldsymbol{c}$ to the sparse array, nonlinear parameters of a pulse shaped function $\boldsymbol{b}$ for each component, and linear parameters $\boldsymbol{p}$ of each component to produce the function:

$$\begin{aligned} y(\boldsymbol{\theta})_k &= \sum_{i=1}^{m} w_i f_i \\ &= w_1[p_{11}(d(x_1) - c_{11}) + p_{12}(d(x_2) - c_{12}) + \cdots + p_{1n}(d(x_n) - c_{1n}) + p_{1(n+1)}) \\ &\quad + w_2[p_{21}(d(x_1) - c_{21}) + p_{22}(d(x_2) - c_{22}) + \cdots + p_{2n}(d(x_n) - c_{2n}) + p_{2.n+1}) + \cdots \\ &\quad + w_m[p_{m1}(d(x_1) - c_{m1}) + p_{m2}(d(x_2) - c_{m2}) + \cdots + p_{mn}(d(x_n) - c_{mn}) \\ &\quad + p_{m(n+1)}) \end{aligned} \quad (4)$$

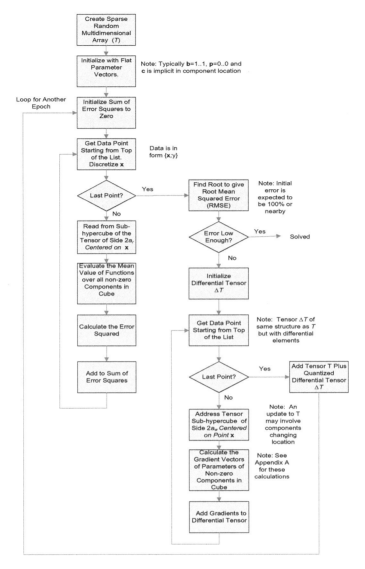

Fig. 2 Discrete neuro-fuzzy tensor calculus

$$w_i = \prod_{j=1}^{n} w_{ij} = \prod_{j=1}^{n} \frac{1}{1 + \beta \left| \frac{d(x_j) - c_{ij}}{a_r} \right|^{2b_{ij}}} \quad (5)$$

where

$\mathbf{c}_i = [c_{i1}, c_{i2}, \ldots, c_{in}]^{\mathrm{T}}$, $\mathbf{b}_i = [b_{i1}, b_{i2}, \ldots, b_{in}]^{\mathrm{T}}$, $\mathbf{p}_i = [p_{i1}, p_{i2}, \ldots, p_{in+1}]^{\mathrm{T}}$ and $d(x_i)$ is an input x_i of $\mathbf{x}$ mapped into an index to the array. The error function to be minimized is

$$e([\mathbf{c}_1..\mathbf{c}_m], [\mathbf{b}_1..\mathbf{b}_m], [\mathbf{p}_1..\mathbf{p}_m]) = (y([\mathbf{c}_1..\mathbf{c}_m], [\mathbf{b}_1..\mathbf{b}_m], [\mathbf{p}_1..\mathbf{p}_m]) - \hat{y})^2 + \varepsilon$$

where $\hat{y}$ is the field data. If $\beta \gg 1$, then b can be adjusted during training without affecting the width of the function w_i, which is a requirement for the tensor field model to work. The appearance of w_i is that of a multidimensional pulse or spike (depending on the value of b) and the reason for the descriptor 'discrete.' The width of the pulse is determined by a and does not change as b changes.

Now, taking the partial derivative of $e_k(\theta)$ with respect to c_{ij}, we have

$$\frac{\partial e_k}{\partial c_{ij}} = 2(y(\theta)_k - \hat{y}_k)\left[f_i \frac{\partial w_i}{\partial c_{ij}} + w_i \frac{\partial f_i}{\partial c_{ij}} \right] \quad (6)$$

Given

$$f_i = p_{i1}(d(x_1) - c_{i1}) + p_{i2}(d(x_2) - c_{i2}) + \cdots + p_{in}(d(x_n) - c_{in}) + p_{i(n+1)} \quad (7)$$

$$\frac{\partial f_i}{\partial c_{ij}} = -p_{ij}, \quad \frac{\partial w_i}{\partial c_{ij}} = \begin{cases} \frac{2b_{ij}}{d(x_j) - c_{ij}} w_{ij}(1 - w_{ij}) & x_j \neq c_{ij} \\ 0 & x_j = c_{ij} \end{cases} \quad (8)$$

$$\frac{\partial e_k}{\partial b_{ij}} = 2(y(\theta)_k - \hat{y}_k) f_i \frac{\partial w_i}{\partial b_{ij}} \quad (9)$$

$$\frac{\partial w_i}{\partial b_{ij}} = \begin{cases} -2 \ln \left| \frac{d(x_j) - c_{ij}}{a_r} \right| w_{ij}(1 - w_{ij}) & x_j \neq c_{ij} \\ 0 & x_j = c_{ij} \end{cases} \quad (10)$$

$$\frac{\partial e_k}{\partial p_{ij}} = 2(y(\theta)_k - \hat{y}_k) w_i \frac{\partial f_i}{\partial p_{ij}}, \quad \frac{\partial f_i}{\partial p_{ij}} = \begin{cases} (d(x_j) - c_{ij}) & j \leq n \\ 1 & j = n+1 \end{cases} \quad (11)$$

$$\nabla e_k(\theta) \equiv \left[\frac{\partial e}{\partial c_{11}}, \frac{\partial e}{\partial c_{12}}, \ldots, \frac{\partial e}{\partial c_{1n}}, \frac{\partial e}{\partial c_{21}}, \frac{\partial e}{\partial c_{22}}, \ldots, \frac{\partial e}{\partial c_{2n}}, \ldots, \frac{\partial e}{\partial c_{m1}}, \frac{\partial e}{\partial c_{m2}}, \ldots, \frac{\partial e}{\partial c_{mn}} \right.$$
$$\frac{\partial e}{\partial b_{11}}, \frac{\partial e}{\partial b_{12}}, \ldots, \frac{\partial e}{\partial b_{1n}}, \frac{\partial e}{\partial b_{21}}, \frac{\partial e}{\partial b_{22}}, \ldots, \frac{\partial e}{\partial b_{2n}}, \ldots, \frac{\partial e}{\partial b_{m1}}, \frac{\partial e}{\partial b_{m2}}, \ldots, \frac{\partial e}{\partial b_{mn}}$$
$$\left. \frac{\partial e}{\partial p_{11}}, \frac{\partial e}{\partial p_{12}}, \ldots, \frac{\partial e}{\partial p_{1,n+1}}, \frac{\partial e}{\partial p_{21}}, \frac{\partial e}{\partial p_{22}}, \ldots, \frac{\partial e}{\partial p_{2,n+1}}, \ldots, \frac{\partial e}{\partial p_{m1}}, \frac{\partial e}{\partial p_{m2}}, \ldots, \frac{\partial e}{\partial p_{m,n+1}} \right]^T \quad (12)$$

Batch learning is achieved by calculating the gradient over all M points

$$\nabla e(\theta^i) = \sum_{k=1}^{M} \nabla e_k(\theta^i) \quad (13)$$

$$\theta^{i+1} = \theta^i - f(\nabla e(\theta^i)), \quad f(\nabla e_k(\theta^i))^T = [\text{sgn}(\nabla e(\theta^i)_{\theta=c}) \cdot q_c \quad \eta \nabla e_k(\theta^i)_{\theta \neq c}] \quad (14)$$

where

$q_c = d^{-1}(i+1) - d^{-1}(i)$ and η is a learning rate for parameters not quantized and $d : \mathbf{x} \rightarrow \mathbf{i}, \mathbf{i} = [i_1, i_2, \ldots, i_n], \quad i_i \in [1, 2, .., N]$ are tensor indices.

These calculations are looped for several epochs $i = 1, \ldots, N$ giving θ^N

$$\theta^N \rightarrow \theta^*, \quad N \rightarrow \infty \tag{15}$$

4 Earthquake Reconstruction Case Study: Data Analysis and Processing

On December 26, 2003, at 5:26 A.M. a 6.5 magnitude earthquake devastated the city of Bam, located in Kerman province in south-east Iran. The Bam earthquake killed more than 27,000 and left 75,000 people homeless. The city's infrastructure, including the water supply, power, communication, health-care services, government buildings, main roads and the only airport, was crippled. The 2400-year old castle of Arg-e-Bam, the biggest mud-brick complex in the world, was totally destroyed (Fig. 3). Data about the reconstruction process are available in tabular form involving 75 different variables from columns 3 to 78 of the table (the first three columns not used) including, for example, the numbers of engineers and technicians and the uses of various building materials and other resources. The structure of the data is shown in Fig. 4.

4.1 Data Analysis

The columns of the earthquake reconstruction data are shown graphically in Fig. 3. The data exhibit strong relationships among several variables; for example, the absolute correlation coefficient of column 13 (engineers) and 14 (technicians) was 0.7921,

Fig. 3 The castle of Arge-Bam, Iran **a** before the earthquake and **b** after earthquake

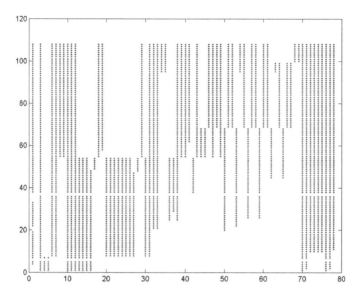

Fig. 4 The distribution of data among the 78 columns and 108 rows in the Bam reconstruction data. The largest contiguous block is only about 8 columns wide and 100 rows long

indicating a strong correlation between the engineers and technicians working in the region directly impacted by the earthquake affected. Vectors of IID random numbers of the same length had a correlation coefficient of 0.0381 (the higher the absolute correlation coefficient, the better the relationship between columns). Analysis of the correlations among other variables was performed.

It is assumed $y = f(\mathbf{x}, \boldsymbol{\theta}) + \varepsilon$ where the scalar y can be any variable, the vector $\mathbf{x}$ consisting of all other variables (the other 75 columns) and ε is noise. The function $y = f(\mathbf{x}, \boldsymbol{\theta}) + \varepsilon$ can be visualized as a surface in n-dimensional space, where n is the length of vector x. The analysis begins by establishing a relationship between the quantity of steel products distributed after the earthquake (column 78) as a function of all other 75 input variables (provided in columns 3 to 77). From this point, forward input $\mathbf{x}$ refers to data in columns 3 to 77, while y refers to column 78 (the quantity of steel products distributed).

Let the output variable y represent the predicted quantity of steel distributed (column 78) as a function of all other variables given in columns 3 to 77. The modeling surface is then constructed from a finite set of data points, each one being a vector x (106 data points). Finally, the model can then be applied to determine the output variable (in this case the quantity of steel product distributed) for elements not within the set of 106 data points for which the network was trained.

4.2 Discrete Neuro-fuzzy Data Processing

In modeling the surface, post-training array components may (or may not) cluster together. We define knowledge about the surface to be "superficial knowledge." If the extracts do cluster, then this produces "semantic knowledge" (a language code). Semantics are a form of knowledge compression, i.e., it represents what could be an extremely complex scenario very concisely—although not precisely. The centers of extract clusters are found using fuzzy clustering and determine the nature of the semantic code (i.e., the rules).

All data were linearly scaled to give values between 0 and 1.0. Input data were further scaled to yield integers between 20 and 80 which serve as array indices. A sparse tensor (multidimensional) array of 106 sparse extracts (the same as the number of lines of data) was used. The tensor array was square with dimensionality 75 and size 100, which gives $100^{75} = 10^{150}$ elements. Only 106 of those elements are nonzero. Extracts had settings $\beta = 100$ with pulse widths $a = 30$, and they were all centered on the positions of the data that were converted to indices of the array. That is $\mathbf{c}_i = \mathbf{x}_i \in [20, 80]^{75}$ for all 106 of i. Defined for each extract in the array is a linear hyper-plane. These hyperplanes were all initially flat, i.e., $p_i = 0$, $i = 1.75$ which are the angles with the horizontal. The heights of the planes above the origin were initially the output of the function (column 78) scaled to between 0 and 1.0 as indicated, i.e., $p_{76} = y$.

Initial values of the parameters determining the shape of the pulses were $b_i = 8$, $i = 1.75$. These values of $\mathbf{b}$ and $\beta = 100$, $a = 30$ produce fixed size rectangular multidimensional pulse shapes for the weightings of extracts. The values $a = 30$ relative to the size of the dimensions of the space at 100 may give the impression that these pulses are large. However, an interesting peculiarity of such high-order spaces is that relative to the size of the space these pulses are very small. Kanerva [4] provides some insight into this phenomenon. This specific initial value for elements of $\mathbf{b}$ is chosen arbitrarily.

The tensor array is now ready to be trained with the data. The very specialized stochastic quantized gradient descent method was used with quantization steps for array location being $q_c = 1$ (one index), quantization steps for slopes of hyperplanes $q_{p1} = 0.000025\pi$ rad, the height of the centers of hyperplanes $q_{p2} = 0.05$ and the shape of pulses $q_b = 0.04$. This means that for each epoch of training these values can only change by the stated quanta. Thus, for example, the location of extracts can only change by one index in any dimension for each epoch.

5 Training Errors and Natural Rules

Training was allowed for 100 epochs with a starting error of 4.1337×10^3 and a final error of 3.8265 (Fig. 5). This error is the total sum of squares over all data points. A value that would give the expected error for any single point would be

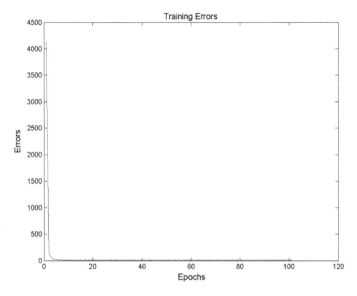

Fig. 5 Training error reduction

the root mean squared value, which after training is 0.19 or 14,618 tons of steel products distributed. This might seem large but can be reduced further arbitrarily by training for longer periods (more than 100 epochs). It is important to highlight the performance of the system with data that were not been previously used in training (i.e., checking data). If checking errors are comparable with the training errors, then a process of prediction or generalization has occurred. Extract locations **c** changed from their initial positions as did the **p** and **b** parameters. The reduction in error indicates that the relationship between input and output is not one of random noise (Fig. 5).

5.1 Error Checking

One may consider separating the data into two equal-sized training and checking sets. However, the aforementioned approach is preferred since the paucity of data may mean that removing the information from the checking set could lead to a lack of sufficient information for model identification. Hence, the proposed approach checks for model accuracy while ensuring enough variation to identify the model (minimizing the risk of contributing to in ergodic data). The average root mean squared checking error for 30 random points was found to be 0.1296 or an error of 9972 tons of steel product distributed. One may notice that the checking error is smaller than the training error. This is consistent with the behavior of low-order problems/systems. One must bear in mind that this prediction is for data that were not used in the training and the predicted steel product distribution given for all of the other 75 variables (columns 3 to 77) has an error comparable with the training error.

5.2 Natural Rules: Extract Clustering

In the discrete neuro-fuzzy approach, no assumptions about knowledge to be discovered are made although there are upper limits due to the number of rules that can be encoded. Rules are found if extracts cluster in the sparse array which can be easily demonstrated graphically for low-order problems. For higher order problems, their dimensionality makes it hard to graphically visualize the clustering of extracts place so this performed computationally. Clustering extracts as proposed in the theory are indicative of the presence of natural rules (that can potentially be encoded into words) about the relationship between the quantity of steel products distributed and other variables. The fuzzy C-means clustering MATLAB fuzzy toolbox function is used to identify clusters. This function (fcm) is run several times with two parameters: **c** extract locations in the array and varying numbers of possible clusters. An optimal number of clusters can be inferred from this algorithm based on a low residual for a given number of clusters. Once natural rules about the earthquake reconstruction process are uncovered from the clustering extracts, then the rule centers can be found.

6 Conclusions and Insights

Natural disasters give rise to complex, self-organizing and ill-structured short-term response and long recovery efforts. After a disaster strikes, the resulting group decision and negotiation processes are characterized by extreme instability, high decision stakes, large-world uncertainty, severe time constraints, multi-faceted negotiations and dynamic, value-laden trade-offs. It is shown that discrete knowledge processing constitutes a robust and efficient soft computing approach for the challenges facing decision makers from the government, not-for profit and private sectors. Specifically, discrete knowledge models allow for data complexity and offer the possibility of knowledge discovery pertaining to wide range of political, economic, environmental and social factors. This enables for more advanced reasoning about the behavior of the disaster system. The proposed discrete knowledge processing model was implemented to capture the reconstruction process after a major earthquake in Iran. On December 26, 2003, at 5:26 A.M. a 6.5 magnitude earthquake devastated the city of Bam, located in Kerman province in south-east Iran. The Bam earthquake killed more than 27,000 and left 75,000 people homeless. The city's infrastructure, including the water supply, power, communication, health-care services, government buildings, main roads and the only airport, were crippled. Training and validations of earthquake recovery data sets were used to successfully establish the relationships among key economic, political and social variables using the discrete knowledge processing approach. The errors for training and checking data were shown to be low after training using the discrete neuro-fuzzy model. This indicates a robust model of the reconstruction process that was obtained and can help to guide future reconstruction efforts.

An innovative fuzzy-neural computing approach has been put forth for addressing the challenges faced by decision makers after a natural disaster. By capturing the complexity and uncertainty of the disaster management process, the discrete group neuro-fuzzy knowledge processing model is used to capture the reconstruction process after a real-world major earthquake. Training and validation of the earthquake recovery data sets establish the key relationships among economic, political, engineering, environmental and social variables, thereby promoting knowledge discovery and providing insights for improved disaster recovery models to assist with future recovery processes.

In summary, while classic neuro-fuzzy systems involve the a priori selection of rules and order reduction (principal components, subtractive clustering, etc.) is often used to reduce the number of rules, the proposed discrete knowledge processing approach promotes knowledge discovery (rather than filling modeling into fit into a specific, preconceived template) and determines the appropriate number of rules (rather than relying on the training limitations of conventional neuro-fuzzy methods).

References

1. R.E. Bellman, *Adaptive Control Processes* (Princeton University Press, Princeton, 1961)
2. W.E. Combs, J.E. Andrews, Combinatorial rule explosion eliminated by a fuzzy rule configuration. IEEE Trans. Fuzzy Syst. **6**(1), 1–11 (1998)
3. S.R. Jang, C.T. Sun, Functional equivalence between radial basis function networks and fuzzy inference systems. IEEE Trans. Neural Netw. **4**(1), 156–159 (1993)
4. P. Kanerva, *Sparse Distributed Memory* (M.I.T. Press, Cambridge, 1998)
5. L.D. Keil, Chaos Theory and Disaster Response Management: Lessons for Managing Periods of Extreme Instability. What Disaster Response Management Can Learn From Chaos Theory (1995), p. 3
6. S. Lele, B. Golden, K. Ozga, E. Wasil, Clustering rules using empirical similarity of support sets, in *Fourth International Conference on Discovery Science*, Lecture Notes in computer Science 2262 (Springer, 2001), pp. 447–451
7. A. Lendasse, J. Lee, E. De Bodt, V. Wertz, M. Verleysen, Dimension reduction of technical indicators for the prediction of financial time series—application to the BEL20 market index. Eur. J. Econ. Soc. Syst. **15**(2), 31–48 (2001)
8. V. Michel, D. Francois, The curse of dimensionality in data mining, in *International Work-Conference on Artificial Neural Networks*, Lecture Notes in Computer Science 3512 (Springer, 2005), pp. 758–770
9. H. Rittel, W. Webber, Dilemmas in a general theory of planning. Pol. Sci. **4**, 155–169 (1973)
10. M. Tsutomu, T. Yamakawa, Fuzzy inference on an analog fuzzy chip. IEEE Micro **15**(4), 8–18 (1995)

Author Index

A
Aleemulla Khan, P., 155
Archana Acharya, T., 227
Aruna, S., 77, 173, 221
Asish Vardhan, K., 53
Ashish Kumar, 181

B
Beebi, Shaik Khasim, 163
Bharat, Talasila, 1
Bhattacharyya, Debnath, 23, 45, 155
Byun, Yung-Cheol, 91, 99

C
Chaitanya, Uppuluri, 35
Challa, Narasimham, 13
Chandini, S., 221
Chandra, J. Vijaya, 13
Chen, Ming, 45, 71, 245, 251
Coffman, David, 115

D
Das, Rudra Pratap, 181
Deshpande, Neha, 105

F
Fung, Carol, 115

G
Gelogo, Yvette, 71
Golagani, Prasanna Priya, 163
Gousia Begam, S. K., 173

H
Hazra, Debapriya, 99
Hazra, Dipankar, 23

J
Jyothi, Rednam S. S., 53

K
Kalluri, Hemantha Kumar, 197, 279
Kavitha, Netala, 145
Krishna, Muddada Murali, 245
Krishna, Sajja Tulasi, 279
Kumar, L. Sarath Chandra, 59

L
Lavanaya, D., 45
Levy, Jason, 115, 263, 287, 303

M
Madhuri, Bonela, 85
Madhusudhan, D., 221
Mahalakshmi, Tummala Sita, 163
Mandhala, Venkata Naresh, 123
Mathew, Jose Alex, 105
Muthusamy, Pachiyannan, 77

N
Naaz, Md. Azma, 123
Nageswara Rao, P. A., 181
Naik, Srinivasa, 173
Nelatur, Naresh, 59

P
Pasupuleti, Sai Kiran, 13
Patnaik, Vurity Sridhar, 133
Patnala, Eswar, 53

R
Ramkumar, Shivram, 115
Rao, N. Thirupathi, 45, 251
Ray, Asmita, 71
Reddy, B. Dinesh, 59
Roy, Sourav, 173

S
Sagar Imambi, S., 123
Sajja, V. Ramakrishna, 197
Sambana, Bosubabu, 133
Sarathchandra Kumar, Labala, 35
Satyanarayana Kalahasthi, V., 245
Satyanarayana, K. V., 251
Shahbazi, Zeinab, 91
Sosnkowski, Alexander, 115
Sowmya, K. B., 105
Srinivasa Naik, K., 77, 221

Sudha, K., 251
Sudhakar, Ch, 85
Sukanya, Y., 181
Suneetha, P., 77

T
Thirupathi Rao, N., 53, 85, 133, 155

U
Umadevi, Viyapu, 181

V
VanKara, JayaVani, 245
Vayelapelli, Mamatha, 145
Veda Upasan, P., 227
Venkatra Krishna Kishore, Kolli, 205
Venkatramaphanikumar, S., 205

Y
Yaswanth Kumar, B., 205